DELTAS OF THE WORLD

This volume is part of a series of volumes on Coastlines of the World.
The papers included in the volume are to be presented at Coastal Zone '93

Volume Editor Robert Kay
Series Editor Orville T. Magoon

Published by the
American Society of Civil Engineers
345 East 47th Street
New York, New York 10017-2398

ABSTRACT

This proceeding, *Deltas of the World*, contains papers presented at COASTAL ZONE 93, the eighth symposium on Coastal and Ocean Management held in New Orleans, Louisiana, July 19-23 1993. This volume is part the continuing series of volumes on Coastlines of the World. Some of the topics covered include environmental considerations, engineering and science; data gathering, and monitoring, legal, regulatory, and political aspects of coastal management planning, conservation and development and public information and citizen participation. This volume provides the professionals, decision-makers, and general public with a broad understanding of these subjects as they relate to the *Deltas of the World*.

Library of Congress Cataloging-in-Publication Data

Deltas of the World / volume editor, Robert Kay
p.cm.— (Coastlines of the world)
Includes indexes.
ISBN 0-87262-962-7
1. Deltas—Congresses. I. Kay Robert (Robert C.) II. American Society of Civil Engineers. III. Series.
GB591.D45 1993
551.4'56—dc20 93-14141
CIP

Library of Congress Catalog Card No: 93-14141
ISBN 0-87262-962-7
Manufactured in the United States of America.

Cover photo by Lambert Studios

FOREWORD

Coastal Zone '93, is the eighth in a series of multidisciplinary biennial symposia on comprehensive coastal and ocean management. Professionals, citizens and decision makers met for five days in New Orleans, Louisiana, to exchange information and views on matters ranging from regional to international scope and interest. This year's theme was entitled "Healing The Coast," emphasized a recurrent focus on practical coastal problem solving.

Sponsors and affiliates included the American Shore and Beach Preservation Association, American Society of Civil Engineers (ASCE), Coastal Zone Foundation, Department of Commerce, National Oceanic and Atmospheric Administration, as well as many other organizations (see title page). The range of sponsorship hints at the diversity of those attending the Coastal Zone '93 Symposium. The presence of these diverse viewpoints will surely stimulate improved coastal and ocean management through the best of current knowledge and cooperation.

This volume of the Coastlines of the World series is included as part of the Coastal Zone '93 Conference. The purpose of this special regional volume is to focus on the coastline and coastal zone management of the *Deltas of the World*.

Each volume of the Coastlines of the World series has one or more guest volume editors representing the particular geographical or topical area of interest.

All papers have been accepted for publication by the Volume Editors. All papers are eligible for discussion in the Journal of Waterway, Port, Coastal, and Ocean Engineering, ASCE.

A ninth conference is now being planned to maintain this dialogue and information exchange. Information is available by contacting the Coastal Zone Foundation, P.O. Box 279, Middletown, California 95461, U.S.A.

Orville T. Magoon
Coastlines of the World
Series Editor

CONTENTS

Holocene Evolution of Deltas on the East Coast of India

Kakani Nageswara Rao* and Noboru Sadakata**

Abstract

Disposition of beach ridges vis-a-vis paleo river courses has helped in recognizing several abandoned delta lobes and reconstructing the stages in the progradation of various deltas on the east coast of India. The sea level was more than 100 m below the present level, around 13 ka, in this region, and the maximum Holocene transgression has risen to +5 m around 6 ka, from where it has fallen to the present level with at least 3 relatively stationary levels in between, as can be inferred from the orientation of different sets of beach ridges. The delta progradation, at present, is mostly by the growth of spits and lagoon-filling.

Introduction

The east coast of India, bordering the Bay of Bengal is predominantly a depositional coast with many rivers building deltas at regular intervals along its entire length, except over a stretch of about 370 km on both sides of Visakhapatnam city where several rock promontories jut into the sea as headlands with associated erosional forms. The rates of progradation, however, in different sections of this 2,300 km long coastline is variable with deltaic sections exhibiting rapid advancement, while the inter-deltaic stretches somewhat lag behind.

*Senior Lecturer, Department of Geography, Andhra University, Visakhapatnam 530 003, India

**Associate Professor in Geography, Hokkaido University of Education, Hakodate, Hokkaido 040, Japan

Of all the deltas on the east coast, the Mahanadi, Godavari, Krishna, and the Cauvery are the major ones, each having a subaerial extent of about 9,500, 5,250, 4,750, and 7,500 sq. km, respectively. Application of aerial photographs for studying the morphology of Indian deltas, during the last three decades, has brought to focus a variety of landforms which are vital in understanding the nature of the growth of a delta.

All these deltas exhibit, more or less , similar landforms and progradational characteristics. Moreover, it appears from a majority of studies that surficial forms in these deltas are predominantly Holocene in age. Data, though sporadic, from the continental shelf region fringing the east coast, reveal the occurrence of lower sea-levels at the beginning of the Holocene, and the fluctuations thereafter, which has largely controlled the evolution of landforms in the deltas of the region.

The objective of this paper is to discuss the evolution of Holocene landforms of the east coast deltas, especially the four major ones, by drawing inferences mostly from earlier works, and also to estimate the Holocene sea level changes in the region, based on the evidences both on land and offshore parts of the coast, including a few radiometric dates.

Morphology of the Deltas

The deltas on the east coast of India exhibit a variety of landforms, such as ancient channels, natural levees, ancient beach ridges, mangrove swamps, lagoons, and barrier spits. These typical deltaic landforms are recognized better from the aerial photographs and satellite imagery rather than in the field, since most of these features have either very low or no relief, hence, are called 'photogeomorphic' features (Niyogi, 1968). Identification of these landforms, especially the ancient channels and beach ridges, helps in understanding the delta development.

Ancient channels are the paleo river courses. Generally, rivers abandon their courses in favor of new routes, especially in their deltaic reaches. The abandoned segments are subsequently buried under later sediments, and hence are obscured from the direct observation on the ground. However, these are recognized from air photos based on tonal variations in characteristic serpentine forms. Existence of a number of such buried channels on both sides of the present river courses is common to all the east coast deltas. In fact, the lateral limit of each delta is demarcated based on the extent of these earlier distributary channels on both sides of the river.

Ancient beach ridges are low, narrow, elongated and nearly parallel sets of ridges representing the former shorelines. The heights of these ridges may vary from a few centimeters to several meters, and spacing between the individual ridges is also variable. A continued progradation of a coastline leaves behind beach ridges at every stage.

The disposition of the ancient channels vis-a-vis beach ridges indicate the changing patterns of river mouths and coastal configurations through different stages in the progradation of a delta. Based on this general theme, several studies were made to reconstruct the various stages in the growth of the four major deltas. The present study attempts to provide a comprehensive picture of all these deltas with suitably reinterpreting the data and modifying the maps for a better understanding of the evolution of these deltas during the Holocene.

Godavari and Krishna Deltas: The Godavari and Krishna deltas, although are two separate deltas built by two major river systems (in fact, the second and third largest, respectively) in India, have almost coalesced together, with only a narrow strip of inter-deltaic plain in between, which is also prograding into the sea due to accretion of the sediments supplied from both sides. The photogeomorphic maps of Krishna Delta (Nageswara Rao and Vaidyanadhan, 1978), and Godavari Delta (Sambasiva Rao and Vaidyanadhan, 1979a), indicate numerous ancient channels on both sides of the present courses with the delta apices at Vijayawada and Rajahmundry, respectively. Similarly, quite a number of sandy beach ridges separated from one another by silty/clayey swales were demarcated even up to about 35 km inland from the present coast in these deltas. In addition, Nageswara Rao (1985), has identified a number of beach ridges, over as much width, in the area between these deltas. A composite map of these two deltas including the inter-delta area clearly indicate the overall continuity in the orientation of the beach ridges in the region (Fig.1). A maximum elevation of about 7 m above the present mean sea level is associated with the innermost beach ridges, which decreases towards the coast. The heights of these ridges is within 2 m from the local ground level both in Godavari (Sambasiva Rao and Vaidyanadhan, 1979a), and Krishna deltas (Nageswara Rao, 1980).

Based on the disposition of the different sets of beach ridges, 4 strandlines were surmised in these twin deltas. Each strandline indicates a major change in the coastal

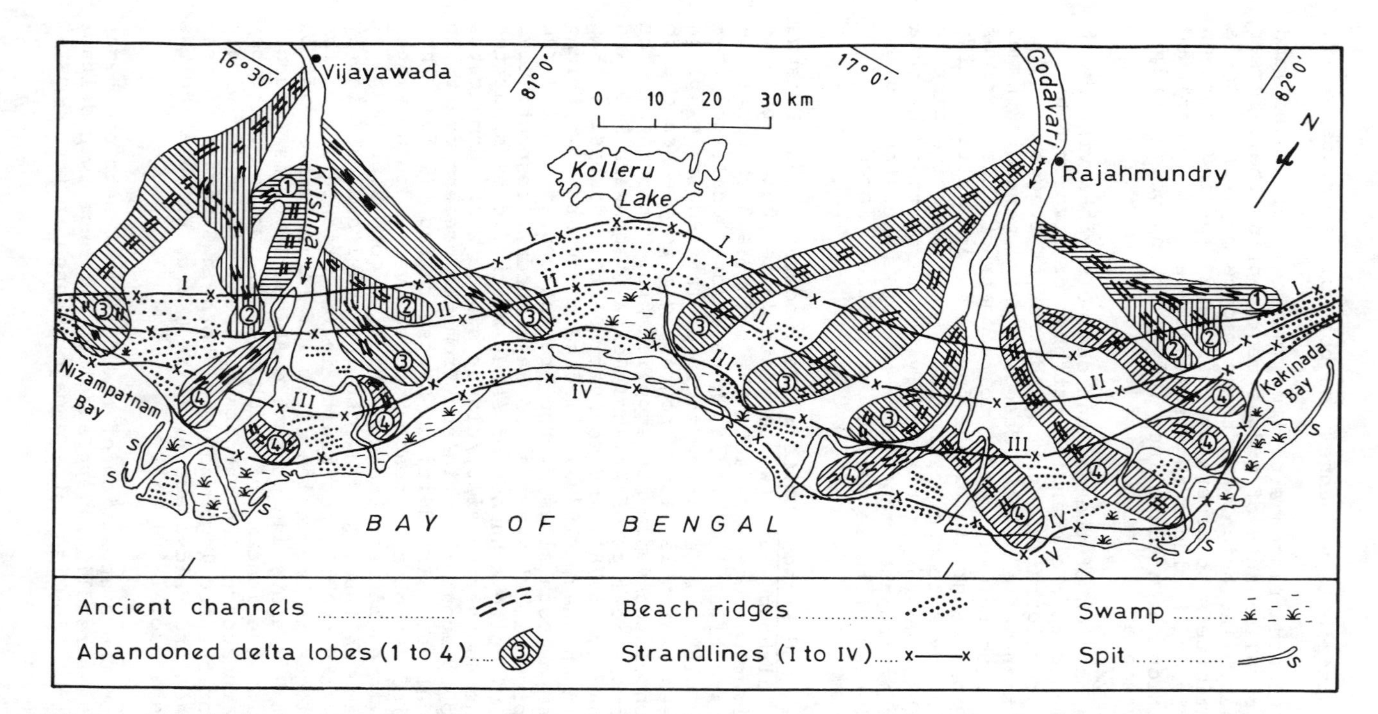

Figure.1 Major photogeomorphic features and stages in the growth of the Krishna (left), and the Godavari (right) deltas. (compiled and modified from Nageswara Rao and Vaidyanadhan, 1978; Sambasiva Rao and Vaidyanadhan, 1979a; and Nageswara Rao, 1985).

configuration. There may be several beach ridges between any two strandlines, but all of them show a similar orientation, indicating the general progradation of the coast at that strandline position. When there is a marked change in the orientation of the coastline, different from the earlier one, probably induced by a change in the location of river mouth, or a drop in sea level, another strandline is said to be initiated, which in turn prograde further creating its own set of parallel beach ridges until another major change in the coastal configuration occurs to initiate yet another strandline.

Similarly, a group of ancient channels which appear to have flown in a certain direction up to a particular strandline, are together considered as one meander belt. Based on the juxtaposition of different meander belts and strandlines, about 11 abandoned delta lobes in Godavari and 9 more in Krishna were identified. Thus, when the coastline was along the strandline I, the rivers flowed into the sea through delta lobes which are numbered as 1. Similarly, when the coastline prograded up to strandline II, the rivers joined the sea through all those distributary mouths which are indicated as delta lobes 2, and so on the delta front of the Godavari and Krishna has advanced through different stages in its evolution.

A logical extension of some of the delta lobes and strandlines into the sea, as shown in Figure 1 (a modification from the earlier studies), is warranted by the orientation of the ridges. Thus, the apparent paleo river mouth position well into the Nizampatnam Bay in the western part of the Krishna delta, and another in the central part of the Godavari front and the possible continuation of the IV strandline through offshore region of the inter-delta area, are suggestive of the shoreface retreat caused by invading sea following the abandonment of a delta lobe and consequent subsidence by sediment compaction, which is commonly associated with a majority of the prograding deltas, as in the case, for instance, of the Mississippi delta (Penland _et al_, 1985).

Thermoluminescence (TL) dating of beach ridge sands in the Godavari delta (Bruckner, 1988) indicated 3.6±0.7 ka, and 3.1±0.6 ka for the ridges at the beginning and ending, respectively, of the II strandline in the western part of the delta, and 3.3±0.65 ka for the ridges midway in the progradation of the same strandline in the eastern part of the delta, indicating that between 3.6 and 3.1 ka the delta has prograded with its front along the II strandline. Also, the ridges at the extreme western and eastern ends of the delta which were demarcated purely based on their orientation as belonging to the II strandline are, in fact, proved to be contemporaneous. Similarly, C-14 dating of fossil shells, peat/wood, and calcrete materials recovered from the beach ridges in the western Krishna delta (Krishna Rao et al, 1990), has indicated, 6.5, 4.5, 3.25, 2.45 and 2.15 ka. Since, specific locations where the dated samples were collected were not shown by those authors, it is tentatively believed here that the first two dates of 6.5 and 4.5 ka perhaps belong to the beginning and ending of the first strandline, while the other three dates could be of the ridges belonging to II, III, and IV strandlines, respectively.

Mahanadi Delta: The Mahanadi delta, the northernmost among the four major deltas is, in fact, a composite delta of the Mahanadi and two more small rivers , namely, Brahmani and Baitarani which interlace together in

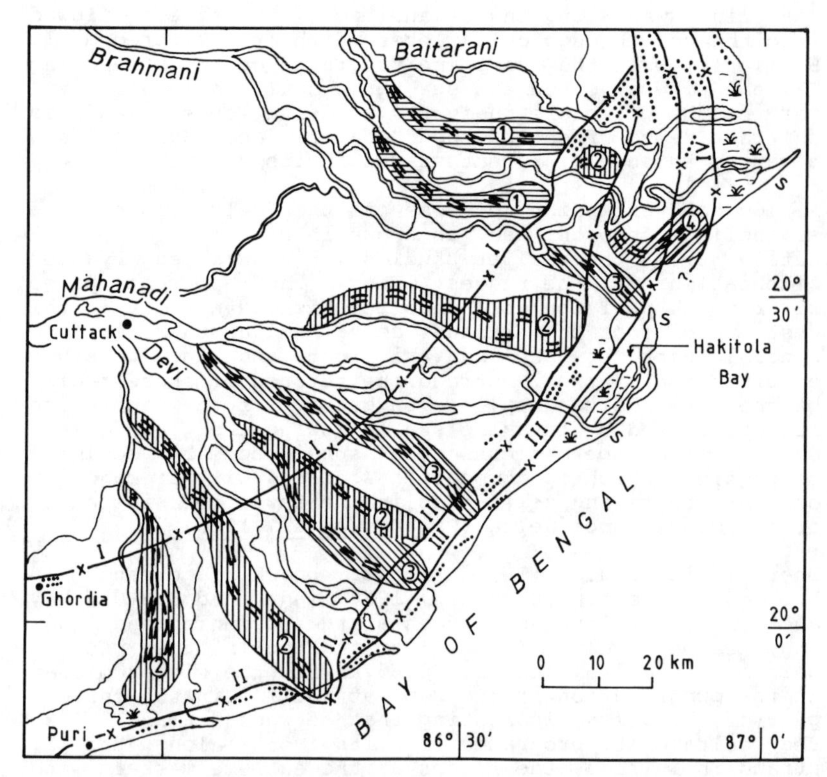

Figure 2. Major photogeomorphic features, and stages in the growth of Mahanadi Delta. See Fig.1 for index. (Modified from Sambasiva Rao et al, 1978, and Meijrink, 1982).

their lower reaches and build their delta at the northern end of the Mahanadi delta. The present Mahanadi with its apex at Cuttack has two major distributaries, namely, Mahanadi and Devi, apart from several minor ones.

Many ancient channels and beach ridges were recognized from the Mahanadi delta too (Sambasiva Rao _et al_, 1978). However, unlike in most other east coast deltas, beach ridges in Mahanadi are poorly preserved, owing to the dominant fluvial action (Meijrink, 1982), which, perhaps, has wiped out most of the ridge deposits. The innermost beach ridges are represented by a small patch of sand mounds at about 35 km from the present coast near Ghordia in the southwestern part of the delta. Their continuity can only be seen in the extreme northeastern end of the delta again at 35 km from the coast. The heights of the ridges in this locality range between 0.3 m and 1.5 m (Mallick _et al_, 1972). In all, 11 abandoned delta lobes and 3 strandlines are surmised in the Mahanadi delta (Fig. 3). The possibility of a fourth strandline associated with one abandoned delta lobe could be seen only in the Brahmani-Baitarani part of this composite delta.

Cauvery Delta: The Cauvery is the southernmost of the four major east coast deltas. With its apex at about 30 km west of Tanjavur, the river flows eastward through its five distributaries. Based on an analysis of surficial and shallow subsurficial deposits, besides air photo study, Meijrink (1971) has discerned a number of ancient channels and beach ridges, and discussed about the stages in the growth of the Holocene delta. Though a later study by Sambasiva Rao (1982) based on the interpretation of some low resolution satellite imagery has missed to discern some of the beach ridges, especially in the northeastern part of the delta, has provided not only the delta growth map, but also useful data from field measurements of beach ridges. The innermost ridge located at 35 km inland has an elevation of about +7 m, with subsequent ridges showing a decrease in elevation towards the coast. Altogether, 3 strandlines and a few delta lobes were surmised in the delta.

In both these studies on Cauvery delta, it was suggested that the earliest coastline was in the southeastern side of the delta and that there was a continuous counterclockwise shift in the delta lobes towards northeast. However, presuming that the innermost beach ridges in the southeastern as well as in the northeastern parts of the delta belong to the same strandline, the Cauvery delta at that point of time could be surmised as a cuspate delta with its mouth somewhere in the present offshore region, as shown in Figure 3. An overextension of the river course by progradation, perhaps, resulted in bifurcation of the distributaries upstream, with new channels flowing southeast as well as northeast while the abandoned lobe might have experienced shoreface retreat.

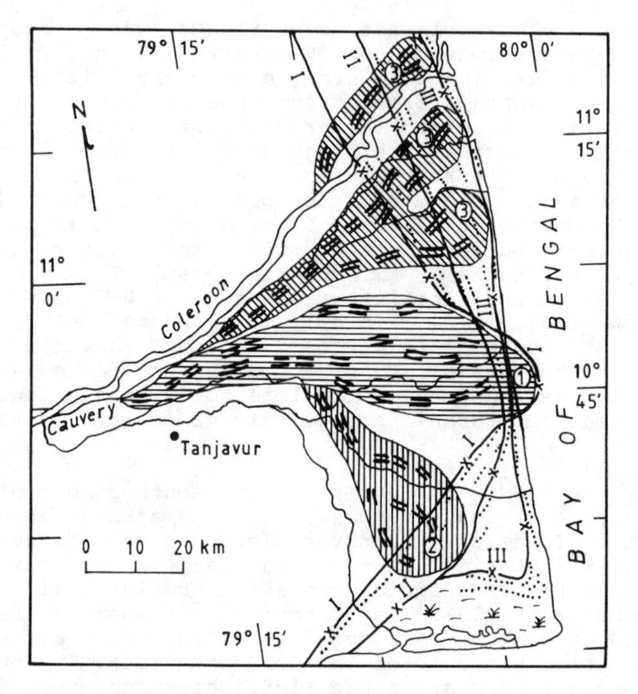

Figure 3. Major photogeomorphic features, and stages in the growth of Cauvery Delta. See Fig.1 for index. (Modified from Meijrink, 1971, and Sambasiva Rao, 1982)

Sadakata (1980), based on bore hole data and a C-14 date (9.92$\pm$0.12 ka), of peat recovered from one of the bore holes at -20 m, has delineated the base of the Holocene sediments at 3 m and 30 m below the present sea level, at about 50 km inland, and near the present coast, respectively.

Photogeomorphic studies from some other deltas along the east coast, have also indicated, more or less, similar forms and conditions. Niyogi (1968) has mapped a series of beach ridges over a width of about 15 km on both sides of the lone distributary of the cuspate Subarnarekha delta, (a small delta with about 850 sq.km area), which marks the northern limit of the east coast where the coast turns eastward and merges into the mighty

Ganges-Brahmaputra delta. Altogether, 4 strandlines were surmised in the Subarnarekha delta. Similarly, about 3 strandlines were inferred at the Penner delta , (another small delta with 1,775 sq.km area, located between Krishna and Cauvery deltas), by Seetharamaiah and Nageswara Rao (1987), and Kameswara Rao (1991). The results of the C-14 dating from the Subarnarekha delta region (Chakravarthy, 1991), has indicated 5.76±0.14 ka for the strandline I, and 2.92±0.16 ka for the strandline II, which are, more or less similar to those obtained in Krishna and Godavari deltas

Present Delta Front Changes

In addition to the long-term changes in the delta development, a few studies have also been made to understanding the present changes at the deltaic coasts. The most common features found along the fronts of the east coast deltas are the spits/barriers and lagoons. A conspicuous spit has grown at the mouth of the northernmost of the 3 distributaries of Godavari. An analysis of maps of different dates has revealed that this 30 km long spit has started growing about 140 years back, perhaps, owing to the increased sediment load due to deforestation in the catchment (Mahadevan and Prasada rao, 1954). The spit has enclosed a large bay called Kakinada Bay, which has been partly filled up and, as a result, mangrove swamp has occupied the area. Further, Sambasiva Rao and Vaidyanadhan (1979b) have mapped a few other smaller spits grown across the distributary mouths of Godavari by the accumulation of the river-borne sediments.

At the Krishna delta front also, a comparative study of maps and aerial photographs, coupled with field mapping by Nageswara Rao and Vaidyanadhan (1979), has revealed the growth of a series of spits, especially at the westernmost distributary. Subsequent filling of the lagoons enclosed by these spits increased the delta area. In fact, the study has indicated that about 30 sq.km area was added to the Krishna delta between 1928 and 1978.

Similar spits are present at the distributary mouths of the Mahanadi also. However, Meijrink (1982) has reported that the Brahmani-Baitarani part of the delta front, and also the large spit at the Hakitola Bay, are facing erosion since 1930. Further, Srinivasan *et al* (1982) while recognizing the erosion of the Hakitola spit, have observed that Mahanadi mouth has shown a northerly deflection several times in the past, by the growth of spits across its mouth from time to time. The present rate of deflection, or in other words, the lengthening of the spit is estimated to be around 200 m per year. Such an overextension of the mouth often leads to the breaching of the spit at its neck during floods/cyclones and the river thus cuts a shorter route to the sea. While another spit may be initiated and grown across the mouth again, the abandoned portion of the channel is filled up and add to the area of the delta. It is apparent that the progradation of the Mahanadi delta during the last few decades is by the growth of a series of spits and channel-fills, while in the case of Krishna and Godavari deltas it is by spits and lagoon-fills (Nageswara Rao, 1988).

Though detailed studies are not made from the Cauvery front, it appears that the entire coastline here indicates no signs of progradation during recent years, except in the northernmost tip around the Coleroon distributary mouth (Ramasamy, 1991). On the other hand, at the Penner delta front, erosion of 5.57 sq.km area has occurred over a 28 km stretch out of the total deltaic coast of 57 km, while growth of spits across the lone river mouth and a partial filling of a small lagoon at the northern end of the delta has resulted in an addition of 6.32 sq.km area between 1918 and 1989 (Srinivasa Rao and Nageswara Rao, 1992). This apparent lack of progradation in the recent decades at some deltas in the region is, perhaps, due to the construction of dams in the river basins, thereby preventing most of the discharge and sediment load from reaching the coast.

Holocene Sea Levels

The available data, even though limited, on the relict sediments and terraces/benches (erosional ?), in some parts of the continental shelf, besides the data on beach ridges, supplemented by a few C-14 dates, reveal the sea level variations during the Holocene in the Indian east coast region.

The occurrence of carbonate reefs at -100 m and -85 m off Visakhapatnam coast (Murthy, 1989), the C-14 dating of which indicated 12.53±0.17, and 10.79±0.17 ka (Mohan Rao *et al*, 1990), and also the identification of terraces/benches of about 2-8 m height around 100 m depth off Visakhapatnam coast (Kukkuteswara Rao and La Fond, 1954, and Murty *et al*, 1992) indicate the sea level positions close to -100 m and -85 m around 12.5 ka and 11 ka, respectively. Moreover, the reported occurrence of a zone of coarse oolitic sand between -100 m and -145 m, continuously along most part of the shelf margin as shown in Figure 4 (Subba Rao, 1964), leads to the presumption of lower sea levels of up to -140 m or more, perhaps at the end of the Pleistocene, since oolites are

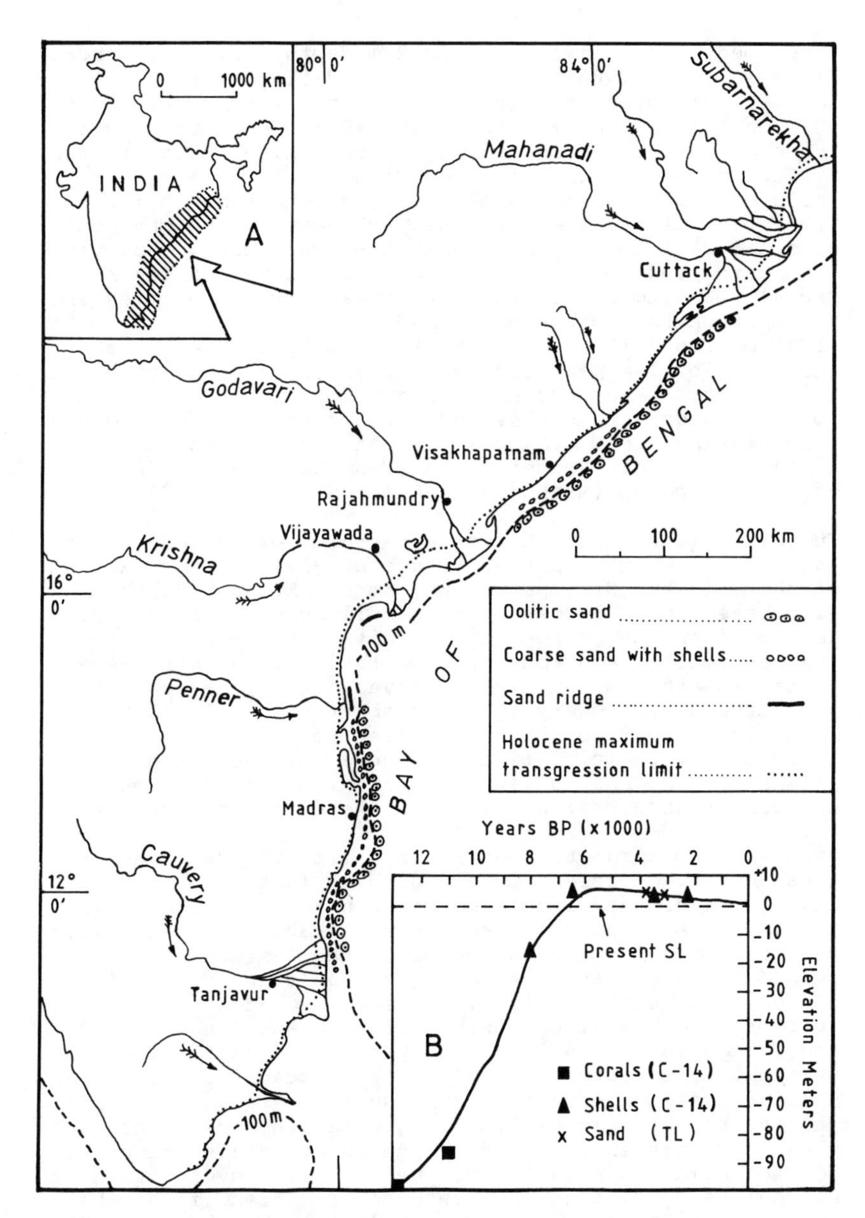

Figure 4. Distribution of relict sediments on the continental margin of the East Coast of India. Inset B in the map shows the preliminary sea level curve for the region.

developed essentially within 2 m depth off the shore (Newell et al, 1960). But the C-14 dating of an oolitic sample collected off Visakhapatnam coast from this zone at about -128 m has indicated 10.8±0.18 ka (Naidu, 1968). These apparently discordant dates which suggest two different elevations to the sea level i.e. -128 m (oolite), and -85 m (coral reef), at the same point of time, i.e. 11 ka, leads to some ambiguity which can only be cleared if more data on ages are available. However, the attempted preliminary sea level curve (inset B in Fig.4) tends to ignore the date given for the oolites and hence suggests a continuous rise in sea level from about -100 m around 12.5 ka to about -85 m by 11 ka. Later, the sea level appears to have risen to around -60 m as evident from the presence of a zone of predominantly coarse sands with shells and foraminifera at that depth off Visakhapatnam (Mahadevan and Poornachandra Rao, 1954), south off Penner mouth (Venkataratnam, 1968), and off Madras coast (Madhusudana Rao and Murty, 1968).

Further, the occurrence of a narrow, elongated relict sand body, akin to a beach ridge at -17 m around 25 km offshore in the Nizampatnam Bay at the Krishna delta, and also based on C-14 dating of a molluskan shell (8.2±0.12 ka) found in that sand body (Srinivasa Rao et al, 1990), and a similar sand body at comparable depth off Penner delta (Seetharamaiah, 1989), reveals that the sea level was perhaps stationary again at about -17 m around 8 ka. From that position, it has risen to +5 m by 6 ka as indicated by the elevations and ages of the beach ridges all along the east coast (see the limit of the maximum Holocene transgression all along the coast in figure 4).

From the maximum transgression limit, the sea has regressed to the present level, with at least two comparatively stationary levels in between - one around 3.5 ka and the other around 2.5 ka, perhaps, at +3.5 m and +1.5 m, respectively, as indicated by the existence of at least three strandlines along the deltaic as well as non-deltaic sections like, for example, at some bayhead (pocket) beaches along Visakhapatnam coast (Bhaskara Rao and Vaidyanadhan, 1975). It may be noted that while each strandline is represented by a set of beach ridges in the deltaic areas, the same is represented by a single ridge in the non-deltaic sections. Whereas the existence of several ridges in each strandline in deltaic sections is due to rapid progradation of the coastline even within a strandline position, the presence of strandlines themselves in the non-deltaic sections as well, probably indicate sea level lowering. The possible forth strandline in some deltaic sections may be purely due to changing location of delta lobes.

Conclusions

The photogeomorphic features like ancient channels and beach ridges found in all the deltas of the east coast of India indicate the abandonment of several delta lobes, and at least 3 different higher sea-stands, than present, during the late Holocene. The tentative Holocene sea level curve for this region is, more or less, similar to the general Holocene curve of Fairbridge (1961) to the extent of presuming higher sea levels than present around 6 ka. As the study reveals, the available data on east coast deltas are not only meager but also conflicting, especially from the offshore region, and hence, the conclusions drawn are bound to be highly conjectural. Detailed morphological and sedimentalogical investigations both on land and offshore regions coupled with intensive data on ages are needed to understand the evolution of some of these major deltas of the world.

References

Bhaskara Rao, V.U. and Vaidyanadhan, R. (1975). "Photogeomorphic studies of coastal features between Visakhapatnam and Pudimadaka". Photonirvachak 3, 43-46.

Brückner, H. (1988). "Indicators for formerly higher sea levels along the east coast of India and on the Andaman islands". Hamburger Geog. Studien 44, 47-72.

Chakravarthy, P. (1991). "Morphostratigraphy of coastal Quaternaries of West Bengal and Subarnarekha delta, Orissa". Ind. J. Earth Sci. 18 (3-4), 219-225.

Fairbridge, R.W.(1961). "Eustatic changes in sea level". Physics and Chemistry of the Earth 4, 99-186.

Kameswara Rao, K.(1991). "Geomorphic features in Penner delta and its evolution". Mem.Geol.Soc.Ind.22, 65-74

Krishna Rao, B. et al (1990)."Sedimentary characteristics of Holocene beach ridges in western Krishna delta". In: Sea level variations and its impact on coastal environment (ed) Victor Raja Manikkam, 133-148.

Kukkuteswara Rao, B. and La Fond, E.C. (1954). " The Profile of continental shelf off Visakhapatnam coast". A.U. Memoirs in Oceanography 1, 78-85.

Madhusudana Rao, Ch. and Murty, P.S.N. (1968)."Studies on the shelf sediments off Madras coast". Bull. Nat. Instt. India 38(1), 442-448.

Mahadevan, C. and Poornachandra Rao, M. (1954). "Study of sea floor sediments of east coast of India". A.U. Memoirs of Oceanography 1, 1-35.

Mahadevan, C. and Prasada Rao, R. (1958). "Causes of the growth of sand spit, north of Godavari confluence". A.U. Memoirs of Oceanography 2, 69-74.

Mallick, S.K. et al (1972) "A comparative study of Quaternary formations in the Baitarani valley, Orissa", Proc.Sem.Geom.Geohyd.Geotech.Ganga basin, 91-104.

Meijrink, A.M.J.(1971)."Reconnaissance of Quaternary geology of Cauvery delta" J. Geol. Soc. Ind.2, 113- 124

------ (1982). "Dynamic geomorphology of Mahanadi delta". ITC Jour. spl. Verstappen Vol., 243-250.

Mohan Rao, K. et al (1990). "Holocene sea levels on Visakhapatnam shelf". Proc. Sem. Conti. Margins of Ind., Waltair, (abstracts), 19-20.

Murthy, K.S.R., (1989). "Seismic stratigraphy of Ongole-Paradip continental shelf". J. Ear. Sci. 16, 47-58.

Murty, P.S.N.et al (1992). "Morphology of the sea floor within the EEZ off Visakhapatnam-Kakinada coast, Bay of Bengal". J. Geol. Soc. Ind. 40, 529-537.

Nageswara Rao, K. (1980) "Landforms and land uses in the Krishna delta" . unpubl. Ph.D. thesis, Andhra Univ.

------ (1985)."Evolution of landforms in the area between Krishna and Godavari deltas".Ind. Geog. J.60, 30-36

------ (1988), "Development of spits and barriers at the deltas". Monograph. on Modern Deltas. 35-38.

------ and Vaidyanadhan, R.(1978)."Geomorphic features of Krishna delta and its evolution". Proc. of Symp. on Morph. Evol. Landforms, Delhi Univ., 120-130.

------ (1979), "Evolution of coastal landforms in Krishna delta front, India". Tran. Instt.Ind.Geog. 1, 25-32.

Naidu, A.S. (1968). "Radiocarbon date of an oolitic sand collected from the shelf off east coast of India". Bull. Nat. Inst. Sci. Ind.38, 467-471.

Newell, N.D. et al (1960). "Bahaman oolitic sand". Jour. Geol. 68, 481-497.

Niyogi, D. (1968). "Morphology and evolution of Subarnarekha delta, India". Tridsskrift Geogr. 67, 230-241.

Penland, S. et al (1985). "Barrier island arcs along Mississippi river deltas". Marine Geol. 63, 197-233.

Ramasamy, S.M. (1991). "A remote sensing study of river deltas of Tamilnadu". Mem. Geol. Soc. Ind.22, 75-89.

Sadakata, N. (1980). "Boring data and fossil pollen analysis in the Cauvery delta, south India". In: Geog. Field Res. in India, Hiroshima Univ. Spl.8. 175-179.

Sambasiva Rao, M. (1982). "Morphology and evolution of Modern Cauvery delta". Trans.Instt.Ind.Geog.4, 67-78

------ et al (1978). "Morphology and evolution of Mahanadi, and Brahmani-Baitarani deltas". Proc. Symp. Morph. Evol. Landforms, Delhi Univ., 241-249.

Sambasiva Rao M. and Vaidyanadhan, R.(1979a). "Morphology and evolution of Godavari delta". Z. Geomorph. 23, 243-255

------ (1979b). "New coastal landforms at the confluence of Godavari river, Ind. J. Ear. Sci. 6, 222-227.

Seetaramaiah, J.(1989)."Studies on modern deltaic sediments of Penner river". unpubl.Ph.D.thesis Andhra Uni.

------ and Nageswara Rao,K. (1987). "Landforms and evolution of Penner delta". J. Landscape Syst.10, 86-91.

Srinivasan, R. et al (1982). "Studies on the morphological changes in the Mahanadi estuary, and Hakitola Barrier island". Photonirvachak 10, 39-44.

Srinivasa Rao, K. and Nageswara Rao, K. (1992). " Coastal forms and processes at the Penner delta front". Pro. 4th Inst. Ind. Geomorph., (Abstracts), p 26.

Srinivasa Rao, P. et al (1990). "Sedimentation and sea levels variations in Nizampatnam Bay, east coast of India". Ind. J. Marine Sci. 19, 261-264.

Subba Rao, M. (1964). "Distribution of calcium carbonate in the shelf sediments off east coast of India". J. Sedimentary Petrology 28(3), 274-285.

Venkataratnam, K. (1968). "Studies on sediments of continental shelf off Visakhapatnam - Pudimadaka, and Pulicat lake - Penner river confluence along east coast of India, Bul. Nat.Inst.Sci. Ind. 38, 470-482.

The Changjiang (Yangtze) River Delta: a Review

WANG Pinxian[1] and SHEN Huanting[2]

Abstract

The Changjaing River delta is the best studied one in China in terms of hydrology, sedimentology and evolution history. The construction of "Pudong New Area" in Shanghai in the 1990es has given a new impulse to the socio-economic development of the delta. Coastal erosion, silting up of waterway in estuary, soil salinization, saltwater intrusion, water pollution, and ground subsidence are the major environmental problems faced by the Changjiang delta now. The growth in economy is accompanied by growing environmental concern, and the gigentic projects of Three Gorge dam and south-to-north water transfer would bring about much more serious problems such as saltwater intrusion, land loss and disturbance of ecosystem to the delta environment. This paper briefly reviews the recent progress in Chinese studies on the Changjiang River delta and discusses its various environmental problems related to its present status, to the future constructions and to the sea level rise caused by global warming.

Introduction

Changjiang (Yangtze) River, with its length of 6,380 km, is the largest river system in China and the third largest in the world. Since the last couple of years, it has become the real focus of econimic development of China and offers a special scientific and socio-economic attraction. Numerous publications have been devoted to its modern environment and evolution history, as well as various problems related to its development (e.g.,Chen et al., 1988; Yan et al., 1987; Wu & Li, 1987; Wang and Li, in press). The present paper is an attempt to synthesize the relevant Chinese literature together with the new data available to us and to discuss the environmental problems faced by the Changjaing delta.

---

[1] Laboratory of Marine Geology, Tongji University, Shanghai 200092, China

[2] Institute of Estuarine and Coastal Research, East China Normal Univrsity, Shanghai 200062, China

The Changjiang River Delta

Physical Settings

The Changjiang River rises in Qinghai-Xizang (Tibetan) Plateau and flows to the East China Sea with a total drop of 6,600m. It drains an area of 1.8 x $10^6 km^2$, about 1/5 of the total area of China. Because its drainage basin is situated in the East Asian monsoon region and distinguished by high precipitation, the water discharge of the Changjiang estuary averages 29,300 cubic metre per second or about 20 times higher than that of the Huanghe (Yellow) River which drainage area is nearly half the size of the Changjiang basin. The total annual runoff of Changjiang averages 924,000 million m^3 with minor year-to-year variations, whereas the water discharge is highly vulnerable to seasonal variations, ranging from 4,620 m^3(in 1979) to 92,600 m^3/sec (in 1954), with 71.7% of the annual total discharge in the flood season (May to October) and 28.3% in the dry season (November to April)(Chen et al., 1988).

The annual average of suspended sediment load in the Changjiang River is estimated at 0.547 kg/m^3 and the total sediment discharge averages 486 million ton per year, ranging from 341 to 678 t/yr and resulting in sediment yield (river load/drainage basin area) of 270 t/yr. The temporal distribution of sediment load is even more concentrated in the flood season amounting to 87% of the annual total, with 21% occuring in July and 0.7% in February. A wealth of sediments settles down when the river debouches from the hilly area of South Jiangsu, forming the modern Changjiang Delta with an area of about $5x10^4$ km^2 including its subaqueous part(Li et al., 1979).

The tidal regime of the Changjiang estuary is semi-diurnal and the tidal range measures 2.66m near the river mouth, reaching its maximum of 4.62m. Averagely, the tidal prism in the estuary reaches 266,300 m^3/sec. which is 8.8 times as much as the average runoff.

Owing to the Coriolis effect,the route of the flood current in the Changjiang estuary deviates northward and that of the ebb current deviates southward. In the retarded current zone between the mainstreams of the flood and ebb currents sediments tend to be deposited, forming shoals and shallows. In consequence, a broad sand bar system has been developed in the Changjiang estuaty, leading to a triple-order bifurcation of the channelized flow which empties into the sea through four outlets: the North Branch, the North Passage, the North Channel and the South Channel (Fig.1). At present, the South Branch is the main passageway of its runoff which in turn is well distributed between the North (36.6-65.3%) and South (34.7-63.4%) Channels, as well as between the North and South Passages (Chen et al., 1988). The current pattern is mirrored by bottom sediments: the South Branch of the estuary is blanketed mainly with sand and the North one with silt (Fig.2).

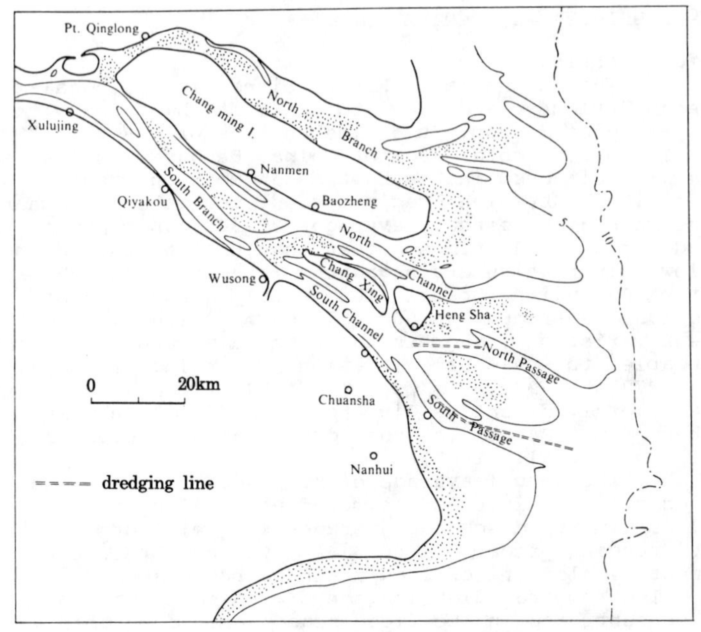

Fig.1 Map of the Changjiang estuary showing its four outlets, isobaths (5m and 10m), and the location of channel dredging

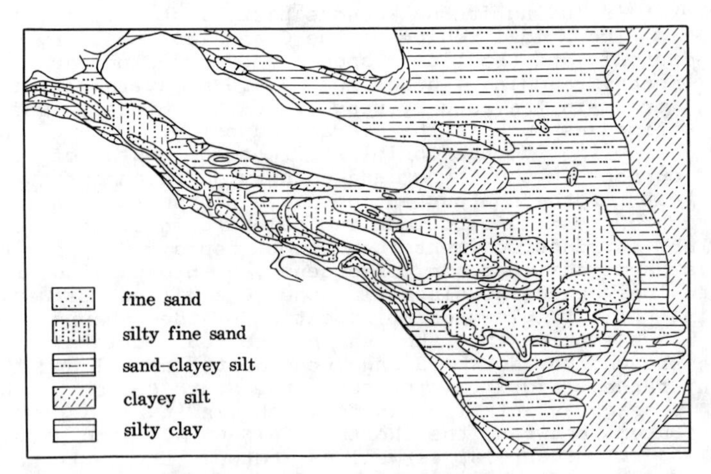

Fig.2 Distribution of sediment types in the Changjiang estuary (modified from Dong and Ding, 1988)

The delta plain of the Changjiang River is averagely 3-5m a.s.l. and very gently tilting eastward with a slope of about one ten thousandth. Geomorphologicaly, the delta plain is limited by hilly areas on the west and southwest, bordered by lacustraine lowlands on the north (Lixiahe lowland) and south (Taihu lacustrine plain).In the main body of the delta channel deposits are developed, while its northern and southern flanks are formed as estuarine-coastal plains (Fig.3).

Economical Settings

The Changjiang delta is economically and technologically the most developed area in China. Its social and economical centre is in Shanghai, one of the three municipalities directly under the Central Government of China, occupying the southeastern part of the delta. With its total area of 6,186 km^2 including the Chongming and other islands and with its population of over 13 million, Shanghai is the largest city in China not only by size of population but also by its significance for the country. Shanghai, for instance, is the largest port in China and contributes about 1/9 of the total industrial output of China.

The rest of the southern part of the delta, the southern Jiangsu, is also witnessing a booming economy boosted by the recent reforms. This is the area renowned for the "richest counties" in the country and for its florishing "village enterprises".

Starting from 1991, the eastern part of Shanghai "Pudong" located on the east bank of the Huangpu River has become the national focus of economical development. "Pudong New Area" becomes the key project in the 1990es' reform and opening of China. This has given a new stimulus to the economic development of the Changjiang delta as a whole.

Evolution Trends

Over a thousand of wells have been drilled in the subaerial part of the Changjiang delta many of which have reached bedrocks. As revealed by drilling, the Quaternary sequence of soft deposits increases its thickness eastward from some 23m to 340m near coast. Although several cycles of alternating sea transgressions and delta formation since early Pleistocene have been reporteded (Wu and Li, 1987), only the Holocene delta is known with certainty.

The Holocene in the Changjiang delta, from 0 to over 30m in thickness, is underlaid by late Pleistocene dark green to brown hard clay of non-marine origin representing whethering surface at the last glaciation. In the northen and southern flanks of the delta plain, the hard clay has been preserved, but eroded by the channels and replaced by channel-filling sands in the main body of the delta (Fig.3). During the postglacial trangression, gray mud rich in foraminifers and other marine microfossils have been accumulated which in turn is overlaid by silty deposits of estuarine and coastal facies resulted from the delta progradation. Judging from the C-14 datings, the present Changjiang delta area was covered by sea water during 10,000 to 7,000 aBP, since then the prograding sequence of the delta was accumulated (Wang and Li, in press).

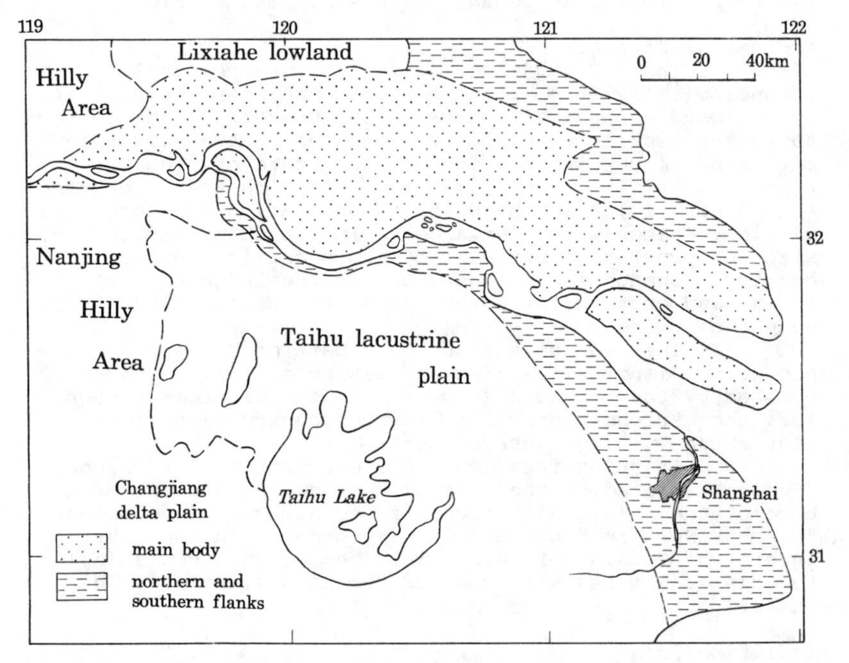

Fig.3 Geomorphological diagram of the Changjiang delta

As shown by archeological, historical, geomorphological and sedimentological data, the Changjiang river had a braod funnel-shaped estuary and later became narrower with the delta progradation. For example, the cheniers on the southern flank of the delta have been dated to 6500 aBP to 800 aBP (Liu et al., 1990). In result of the growth of shoals along the southern bank and the merging of sand bars to the northern bank, the estuary area has lost a half of its original width at its mouth. Meanwhile,the above mentioned assymetric pattern of tidal currents has caused a progressive southern shift of the ebb current route, resulting in a continuous southern migration of the delta system. As seen in Fig.4, the Changjiang estuary has witnessed six stages of evolution since about 7,000 aBP, each with its own sand bar and

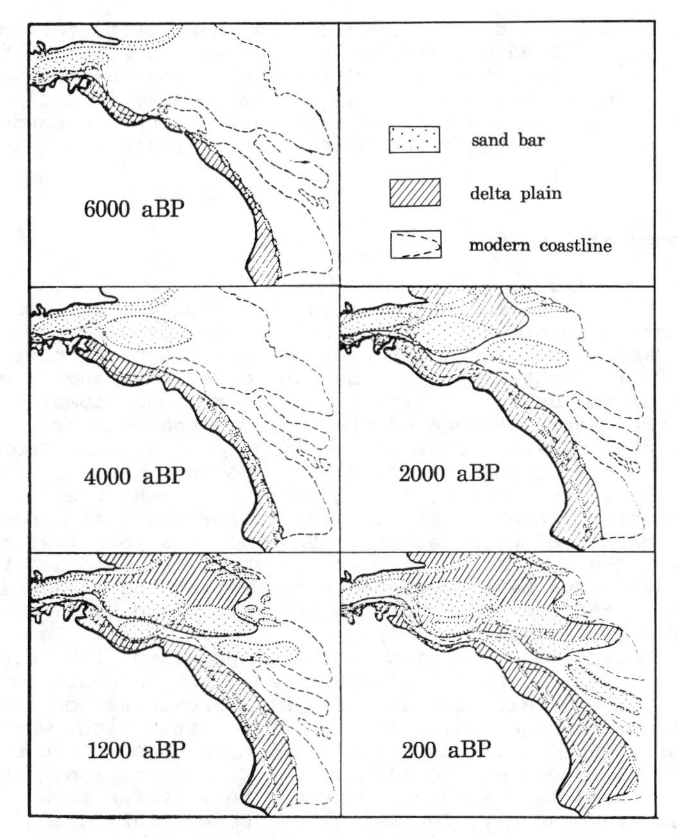

Fig.4 Evolution of the Holocene Changjiang delta showing its first five stages (6,000-200 aBP; the last stage is the modern delta) (modified from Li et al., 1978)

coasts (Li et al., 1979; Xu et al., 1987). However, there is no unanimity in age estimation of the evolution process. Prof.Chen et al.(1984, 1988) have shown that the sediment supply was limited before 2,000aBP and the successive migration of the estuary system and the narrowing of the estuary must have happened within 2000 years.

The same evolution trand has been maintained or even strengtherned at the present time. As the historiccal records show, the shoals of the south bank of the Changjiang estuary prograded seaward at a rate of 25 m/yr, but recently the rate increased to 1 km per 43 m/yr. In the 18th century, the course flow of the Changjiang diverted from the North Branch to the South Branch. Since then the channel of the North Branch has been getting narrow and silted up, and its runoff has decreased to about 25% of the total water discharge of the estuary in 1915 and then to 1-2% by the end of the 1950es, sometimes leading to a backward flow of salt water with suspended load from the North to the South Branch (Chen et al., 1984).

Environmental Problems

Current Problems

The Changjiang delta is facing so many environmental problems that can not be all covered by this review, and only a few major problems are discussed here.

1. Coastal erosion has caused serious land loss problem in the delta. All the over 100 km long coastline along the North Branch of the Changjiang estuary is subject to erosion. About 450 km long coast of the South Branch and the Chongming Island belongs to Shanghai, of which 48% is in erosion, 38% in accretion with the rest being stable. About 63 km (or 14%) of the Shanghai coast is in danger of being eroded. Several tens of cuspate bars, 460 T-shape bars, 30 km of longshore dams and 190 km revetments have been constructed to protect the coast in Shanghai, and over 60 thousand hectars of land have been reclaimed since 1950.

2. The Changjiang estuary provides the sea-entering waterway not only for Shanghai, the largest port in China, but also for all the six provinces of the Changjiang drainage basin. The sand bar system with water depth of about 6 m in the estuary, however, has seriously hampered the use of Changjiang in transportation. For instance, vessels over 5,000 tons had to wait for tide to play in the estuary. Thanks to the successful dredging (Fig.1) a waterway of 7 m dep was finally channeled in 1975. Since then vessels of 10,000 tons can enter in Shanghai Port at any time and full-loaded vessels of 20,000 tons may negotiate the estuary at flood tide. Unfortunately, the dredged channel of the waterway has suffered considerable refill, and the waterway as such operated only for 5 years. The annual volume of dredging for waterway maintenance amounted to over $9x10^6$ m^3 in 1985 and rose to $15x10^6$ m^3 in 1990(Shen et al., 1992).

3. The saltwater intrusion in the Changjaing estuary occurs generally in the dry season (November - April), and the problem is most serious in the North Branch with its minimal runoff. Nevertheless, it can cause grave consequences to the South Branch as well. As an example, the drought in 1978-1979 had led to a great reduction of water discharge in Changjiang and a severe saltwater intrusion from winter 1978 to spring 1979 when the salt water reached 170 km upstream the estuary. The saltwater intrusion lasted 5 monthes and the salt water affected the mouth of the Huangpu River, the last

tributary to Changjiang cutting through Shanghai, for 215 days with the maximum chloride content of 3,950 ppm, bringing about serious damages to industry and agriculture in Shanghai. To prevent the damages, a reservoir was constructed on the bank of the South Branch for storing freshwater with a storage capacity of 10 million m^3.

4. Soil salinization is another environmental problem for the Changjiang delta. Various kinds of salt-affected soil is distributed basically in the coastal zone and along the estuary (Fig.5) with a total area of 3.5 million *mu* or 233 thousand hectares. Soil salinization is more severe in the North Branch where the fresh-water runoff is limited and the tidal effect is strong. This can be illustrated with a comparison between two cities:in Nantong, North Branch, the salt-affected area makes up 42% of the total cultivated area, whereas in Shanghai, South Branch, it accounts only for 10%. Despite of the general trend of desalination in the Changjiang delta, soil salinization is still hindering the agriculture development.

5. Water quality is of greatest concern in the Changjiang delta. Shanghai alone produces over 5 million tons of sewage per day which is drained to the Changjiang estuary or its tributary, the Huangpu river, without treatment. The daily drainage of sewage from the Huangpu River to the Changjiang estuary reaches 3 million tons, being the largest pollution source in the lower reach of Changjiang. While the Changjiang estuary has a fairly great capacity for self-purification because of its high runoff, rapid flow and strong tides, the sewage causes much worse pollution problems to the Huangpu River.

Other environmental problems faced by the Changjiang delta include flooding, solid waste disposal, ground subsidence, etc. The undue utilitation of groundwater has since long given rise to accelerated ground compaction and ground subsidence in Shanghai. The rate of subsidence averaged 0.4 mm/yr within 1910-1919, increased to 24 mm/yr for 1921-1948, 40 mm/yr for 1949-1956 and,finally, 110 mm/yr for 1957-1962. Since 1966, efficient measurements such as water reinjection have been taken to prevent the further sibsidence, and now the subsidence rate is lowered to 3-4 mm/yr. However, ground subsidence is now happpening in other cities of the Changjiang delta where radip economic development has just started.

Problems Related to New Construction Projects

In the course of economic development, China has planned ambitious construction projects which would have serious environmental consequences. Two gigantic projects, the Three Gorge dam and the south-to-north water transfer, are most relevant to the Changjiang delta. Meanwhile, the development of Shanghai Port requires improvement of waterway in the Changjiang estuary. All these are worth a special discussion.

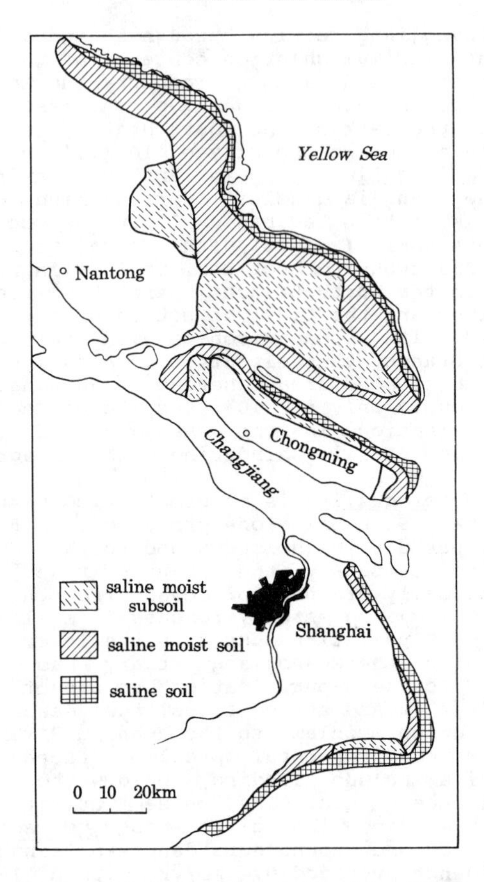

Fig.5 Distribution of salt-affected soil in the Changjiang delta

1.*Shanghai Port* plays an essential role in making Shanghai into a modernized international metropolis and in developing all the Changjiang Basin. To develop the port, both the harbour and the waterway must be improved. For example, by the end of 1993 the construction of four quay-sided 10-thousand-ton class deepwater berths in the Pudong New Area will be completed , and its port handling capacity will reach 2.4 million tons. Site selection of a new port area for Shanghai is undergoing. On the other hand, the existing 7 m deep waterway in the Changjiang estuary is far from sufficient. It has been planned, therefore, to deepen the sea-entering way to 8.0-8.5m toward the year 2000 and further to 12.5 m in 2020. Since the waterway in the Changjiang estuary is unstable under natural conditions, it is necessary to improve the analysis and prediction of the channels' changes for the depth maintenance and further waterway deepening.

2. The *Three Gorge dam* is the greatest hydraulic project planned in China. The dam, 175 m in height and 1983 m in length, will be built in the reach of Changjaing, about 4,500 km from the river head and 1,800 km from the river mouth. The Three Gorgr reservoir is designed for flood protection, electric power generation, navigation and others. It will be 600 km in length, averagely 1.1 km in width and 70 m in depth. Its reservoir capacity will be 39.3 billion m^3 and the installed capacity of hydroelectricity generation is 17.68 million kilowattes. Along with its great socio-economic benefits, the Three Gorge Project will have grave environmental impacts on the Changjaing River and its delta.

Although the reservoir will not significantly change the annual water discharge of Changjaing to the sea, the temporal distribution of runoff within a year will be different. Since the reservoir is planned to draw off water in late May and June, to keep the low water level of 145m from June to September, and to store flood water in October, the monthly runoff in the estuary will be increased for January to May and reduced in October as compared with the present distribution (Table 1).

Table 1 Monthly distribution of Changjaing runoff at Datong Station before and after the Three Gorge dam construction

	present		after dam construction	
month	average runoff for month m^3/s	percentage of annual total %	average runoff for month m^3/s	percentage of annual total %
January	10,630	3.1	11,200	3.3
February	10,710	2.8	11,860	3.2
March	14,420	4.2	15,080	4.4
April	22,260	6.3	21,890	6.3
May	34,690	10.2	38,450	11.3
June	41,060	11.7	41,530	11.9
July	49,730	14.6	49,210	14.5
August	45,480	13.4	45,310	13.4
September	41,210	11.7	40,940	11.7
October	36,010	10.6	30,490	9.0
November	24,880	7.1	24,180	6.9
December	14,710	4.3	13,990	4.1

After the dam construction a large amount of sediment load yielded by the upper and middle reaches will settle down in the reservoir and, hence, the silt concentration in the estuary will decline. At present, the annual mean suspended sediment discharge is estimated at

468 million tons at Datong station, km upstream from the Changjaing mouth. The annual discharge will reduce by 114 milion tons or 23.4% during the first 50 years after the dam construction and will gain its balance toward the 100th year. While the decrease in silt discharge is of benefit to the depth maintainance of waterway in the estuary, it will intensify land loss by accelerating the coast erosion and slowing down the accretion in the delta area. The changes in seasonal variations of runoff resulted from the dam construction will also affect the salt-water intrusion. Compared to the natural conditions, the water discharge at Datong will reduce in October by 32.4% in low flow years, by 20.3% in average years, and by 16.9% in high flow years, while in February it will increase by 24.5%, 19.9% and 5.1%, respectively. In general, the salt water intrusion will maintain longer in October and November with the reduction of freshwater runoff. Because of the regulation of water discharge by the reservoir, the water level in estuary will rise by 13.6cm in February and March, and drop by 21.2cm in October, resulting in corresponding fluctuations of ground water level. The minor fluctuations would not severely affect the general trend of desalination, but might locally retard the process.

3. The south-to-north water transfer project is to transfer water from the Changjiang River Basin to the water-short Huang-Huai-Hai Plain of north China. There are three transfer routes under discussion of which the East Route is from Yangzhou on the northern bank of the Changjiang estuary to the Huanghe (Yellow) River and then to Beijing. According to the project, 1,000m^3/sec is to be tranferred from the Changjiang estuary via the East Route. This is 1/20 of the average annual discharge at Datong station, 1/6 of the minimum dry season discharge, 1/50 of the June-July average and 1/10 of the January average. The transfer of 1,000 m^3/sec from the Changjaing will have little effect on the estuary during the flood season, but in the dry season, especially in low flow years, the impact would be remarkable. When the discharge at Datong drops to less than 16,000 m^3/sec, therefore, water should be transferred prudently, or even better, not at all. The water transfer, amongst others, will will also alter the ecological environment of fishery resources (Shen et al., 1983).

Problems Related to Sea Level Rise

The sea level rise is already a serious problem faced by the Changjiang delta. As shown by instrumental records in Shanghai at the Huangpu River mouth, the sea level shows a clear rising trend in the 30 years from 1955 to 1985, the average rate being 3.2 mm/yr, whereas no rising trend has been recorded for the years 1912-1954. If the rising rate maintains, the sea level will be by 13, 26 and 35 cm higher than now for the years 2030, 2070, 2100, respectively. Taking into account the predicted increasing rate of sea level rise caused by the "greenhouse effect", the threat would be even worse. According to IPCC (1990), the sea could rise 31, 65 or 110 cm by 2100. If the factor of ground subsidence, the sea level might then rise 150 to 200 cm in Shanghai. Since the Shanghai area is mostly lying between 1 and 2 m a.s.l., 13.8% of its total land area would be inundated if sea rises 50 cm, and it reaches 34.7% and 96.1% when sea rises 100 and 200 cm, respectively. Moreover, the rised water level in the Changjiang estuary would also amplify the damages of floods and storms.

Another aspect is saltwater intrusion. As shown by calculations, the 500 ppm isohaline would shift upstream by 16, 32 and 63 km, if the sea rises 50, 100 and 200 cm. Of course, the global warming would also increase precipitation in the delta area which in turn could raise freshwater runoff and partly compensate the effect of saltwater intrusion (Chen, 1992). Nevertheless, the salt water would have disastrous consequences in terms of water supply, soil salinization, ecological environment for fishering and others.

Concluding Remarks

The rapid development of Pudong New Area since the 1990es denotes the regeneration of Shanghai and opens a new era of the Changjiang delta. Newer and newer ambitious projects are under duiscussion, such as the project of sealing North Branch of the Changjiang estuary, the project of large-scale land reclaimation of the delta leading to reach offshor Sheshan Island. However, the enviromnetal aspect should never be ignored while planning new construction, not to mention the existing environmental problems in the delta area like water quality deterioration caused by pollution and saltwater intruion. The approved Three Gorge Project and the proposed south-to-north water transfer project, both will change water discharge in the estuary and leade to serious environmental consequences to the delta. Together with the global warming, all the construction projects are to be carefully investigated for their environmental impact both in quality and quantity, particularly in such a densely populated area as the Changjiang delta.

Acknowledgements Prof.Li Congxian is thanked for the beneficial discussions. Thanks are also due to Ms. Wu Meiying and Mr.Liu Zhiwei for their technical assistance.

REFERENCES

Chen Jiyu, Zhu Huifang, Dong Yongfa and Sun Jiemin, 1984: Development of the Changjiang estuary and its subaqueous delta. In: *Proccedings of Intyernational Symposium on Sedimentation on the Continental Shelf with Special Reference to the East China Sea, April 12-16, 1983, Hangzhou, China*. China Ocean Press, Beijing, 34-51.

Chen Jiyu, Shen Huanting, Yun Caixing et al., 1988: *Processes of Dynamics and Geomorphology of the Changjiang Estuary*, Shanghai Scientific and Technical Publishers, 454p.(in Chinese)

Chen Shenliang, 1992: *Sea Level Rise and its Impact: a Case Study of the Shanghai Area*, Doctoral dissertation of the East China Normal University, 75p.(in Chinese)

Dong Yongfa and Ding Wenjun, 1988: Relationship between the grain size characteristics and hydrodynamics of sedimentation in the Changjiang estuary. In: Chen Jiyu et al., *Processes of Dynamics and Geomorphology of the Changjiang Estuary*, Shanghai Scientific and Technical Publishers, 314-322 (in Chinese)

Li Congxian, Wang Jingtai and Li Ping, 1979: Preliminary study on sedimentary facies and sequence of the Yangtze delta. *ournal of Tung-Chi University*, 1979 (2): 1-14 (in Chinese)

Li Congxian, Guo Xumin, Xu Shiyuan, Wang Jingtai and Li Ping, 1979: The characteristics and distribution of Holocene sand bodies in Chagjiang River delta area. *Acta Oceanologica Sinica*, 1(2): 252-268 (in Chinese)

Liu Cangzi and Cao Ming, 1990: Formation of cheniers and chenier plain of East China. In: *Proceedings of International Symposium on the Coastal Zone*, China Ocean Press, Beijing, 47-61.

Shen Huanting, Mao Zhichang, Gu Guochuan and Xu Pengling, 1983: The effect of south-to-north transfer of saltwater intrusion in the Changjiang estuary. In: A.K.Biswas et al.(Eds.),*Long Distance Water Transfer -- A Chinese Case Stude and International Experiences*, Tycooly International Publ., Dublin, 351-359.

Shen Huanting, Xu Haigen and a Xiangqi, 1992: Study on the improvement od sea-entering waterway in the Changjiang estuary. *China Ocean Engineering*, 6(2): 361-368.

Wang Pinxian and Li Congxian (Eds.)(in press): *Late Quaternary Stratigraphy of the Changjiang Estuary.*

Science Press, Beijing (in Chinese)

Wu Biaoyun and Li Congxian (Eds.), 1987: *Quaternary Geology of the Changjiang Delta*. China Ocean Press, Beijing, 166p.(in Chinese)

Yan Qinshang and Xu Shiyuan (Eds.), 1987: *Recent Yangtze Delta Deposits*, East China Normal University Press, Shanghai, 438p.(in chinese)

Xu Shiyuan, Wang Jingtai and Li Ping, 1987: On evolution stages of the modern Changjiang River delta. In : Yan Qinshang and Xu Shiyuan (Eds.), *Recent Yangtze Delta Deposits*, East China Normal University Press, Shanghai, 264-277 (in Chinese)

CHARACTERISTIC OF THE YELLOW RIVER DELTA

Cheng Guodong*
(Associate professor)

ABSTRACT

The Yellow River Delta is a most unstable delta in the world. The river burst its lowstream bank 1500 times during the last 3000 years, 26 large shifting of the channel as well. The river mouth migrated widely with a distance of 600 km. Several ancient deltas were found along the coastal zone of Bohai Sea and Yellow Sea. The last migration occurred at 1855 and the Modern Yellow River Delta has been constructed since that time. The distributary of the Modern Yellow River Delta shifted 9 times within the 137 years. The coastline of the delta migrated frequently with the evolution of the deltaic lobes. The instability of the Delta is controlled by several factor, such as tectonism of China-plate, sea level change, feature of the receiving ~basin and background of the drainage area.

INTRODUCTION

The Yellow River is one of the long river in the world. It rises from the northern foot of the Bayan Kar Mountains in the Tibet Plateau and passes through the Loess Plateau and empties into the Bohai Sea with a total length of 5464 km and a drainage area of 752 443 km2 (Fig.1).

The Yellow River Delta is a complex delta consisting of a series of ancient and modern deltas located at the western coastal zone of Bohai sea and Yellow Sea in China. It is a most unstable delta in the world. According to the history of China more then 1500 floods occurred in the last 3000 years and every flooding killed hundred thousands people. The river mouth shifted more then 26 times in a distance about 600 km along the coastline. The flooding area extended twenty

---

* Institute of Marine Geology, Ministry of Geology and Mineral Resources , P.O.Box 18, Qingdao, China

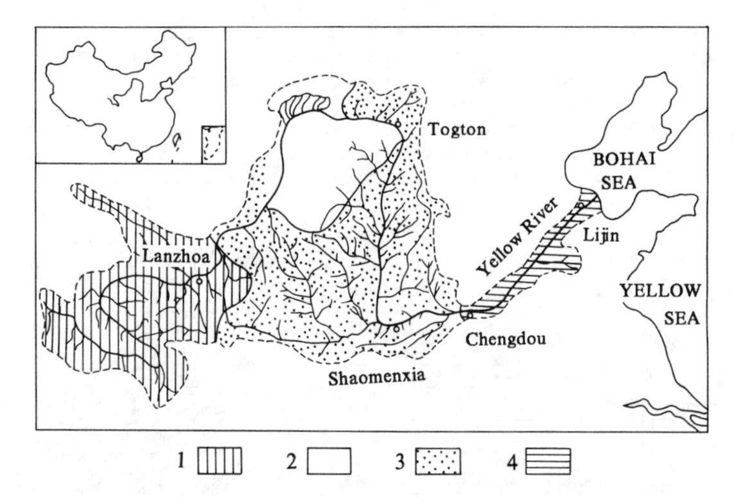

Fig.1 Drainage area of the Yellow River Delta.
1. Upstream drainage area with the elevation of >4000 m. 2-3. Midstream drainage area with the elevation of 2000-1000 m., the area with dots is loess area.
4. Lowstream drainage area with the elevation of <200 m.

five thousand kilometers.

The instability of the Delta is controlled by several factors, such as the tectonism of China-plate, the sea level change, the feature of the receiving basin and the background of the drainage area.

BACKGROUND OF THE YELLOW RIVER DRAINAGE AREA

The geological and geomorphological feature of the Yellow River drainage area is strongly influenced by the activity of China-plate. The subduction of the Pacific plate result in subsidence of the east part of China. The collision between the Indian and Eurasia plates cause the uplift of the west part of China. This topographic form provides an important condition for forming the Yellow River system.

The west high plateau of China is called Qinghai-Tibet Plateau that is a "welding" plate composed of four microplates. It was formed by subduction of the ancient Tethys plate in the Cretaceous Period and Eocene Epoch as a first stage, and uplifted to the elevation of >4000 meters by the collision of the Eurasia and Indian plates after Pleistocene as a second stage. The morphology of the Qinghai-Tibet plateau is divided into three parts: the plateau basin region in the north, the parallel range region in the east and high mountain deep valley region in the southeast. The boundary of these regions is a drainage divide making the water systems towards to the lake basin, Pacific Ocean and Indian Ocean, respectively (Pan and Li, 1990). The Yellow River is come from the parallel range region (Fig.2).

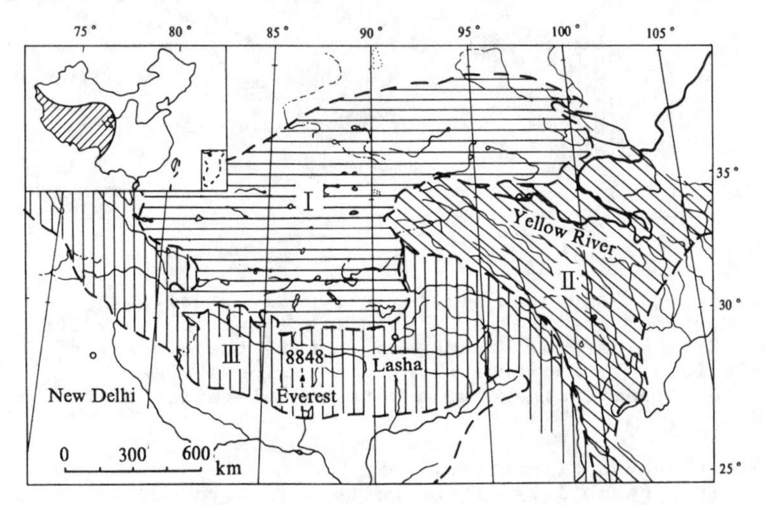

Fig.2 Morphological divisions of the Qinghai-Tibet Plateau. 1. Plateau basin region. 2. Parallel range region. 3. High mountain deep valley region (from Pan Guitang et al,1990)

The east part of the Yellow River drainage area is located at North China subplate which is composed of a series of plate segments. The midstream of the Yellow River is located in one of the segment, the Ordos plateau which is a loess plateau. The thickness of the loess is about two hundred meters. The loess materials were carried here from middle Asia by the west wind drift which was formed when the Qindhai-Tebit Plateau rose up to 4000 meters high. The suspending load of the Yellow River mostly come from the Loess Plateau. The lowstream of Yellow River is located on the another segment, the North China Plain, which was subsiding after Cretaceous and became a wide plain in Quaternary (Fig.3). This provides a extensive place for the migration of the Yellow River channel.

Recently the vertical movement of the China Plate continues uplift with a rate of 10 mm/y in the Qinghai-Tebit Plateau and subsiding with a rate of 4mm/y in the

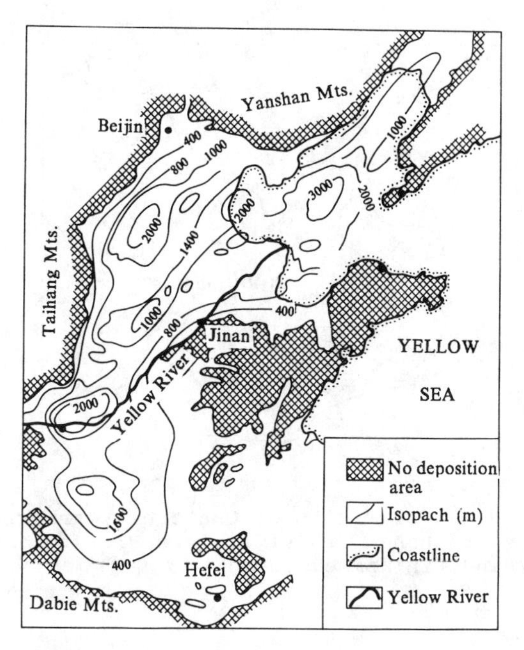

Fig.3 North China Plain and the isopach of upper Tertiary and quaternary.

North China Plain. The yellow River runs eastward and builds the Delta continually.

BACKGROUND OF THE RECEIVING BASIN

The receiving basin of Yellow River, the Bohai Sea, is a part of the North China subplate. During the Quaternary the Bohai Sea became a subsiding center of the North China Plain and is influenced by the sea level change. It is a very shallow sea. Three transgressions were found in the west coast zone after late Pleistocene (Yang & Lin,1991). Those occurred in 75000-127000y.B.P., 23000-31000 y.B.P.and 15000-6000 y B.P. respectively (Fig.4).

The last transgression , the Holocene transgression, arrived to the Bohai Sea at 8000 years B.P. and reached the highest sea level at 6000 years B.P., while the coastline of the Bohai Sea is in the west 200-400 km away from recent coastline. Since that time the Yellow River Delta has been built up along the west coast zone of Bohai Sea and Yellow sea.

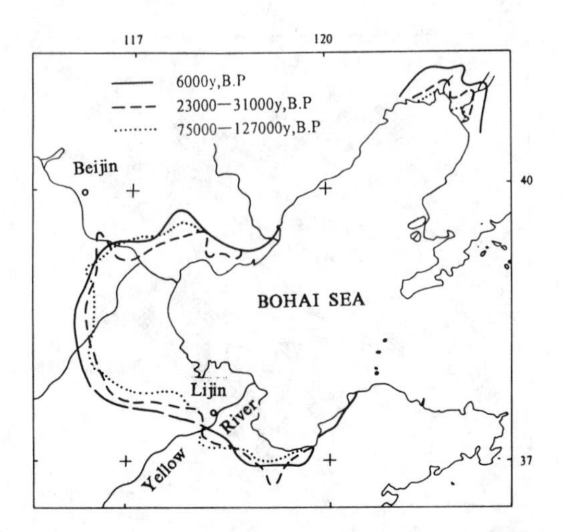

Fig.4 Sketch map of limits of Quaternary transgression on coast area of Bohai Sea after late Pleistocene (from Yang Zigeng and Lin Homao, 1991).

EVOLUTION OF THE YELLOW RIVER

Before the unit Yellow River was formed, there were several independent water systems separated in the modern Yellow River drainage area. The separate water systems, as called pre-river-system, existed there from Cretaceous and linked each other as the Qinghai-Tebit Plateau rose rapidly after Pleistocene.

According to chang (1989) data, several separated water systems occurred in the fault grabens surrounding the boundary of the Ordos plateau in Tertiary. During late Pleistocene Epoch the Sanmenxia lake, one of the independent water system, cut its east bank down due to the uplift of the western part of China. The erosion basis of the Sanmenxia water system declined 40-60 meters and caused the headward erosion. The water system extended to the head direction and connected the upper water system. The separate water systems linked each other gradually from lower to upper, and formed an unit Yellow River System eventually (Fig.5).

CHARACTERISTIC OF THE YELLOW RIVER DELTA

The characteristic of the Yellow River Delta is instability. The earliest Yellow River flows northward into Bohai Sea before 602 B.C..The channel migrated tens times until 1128 A.D .After that the river

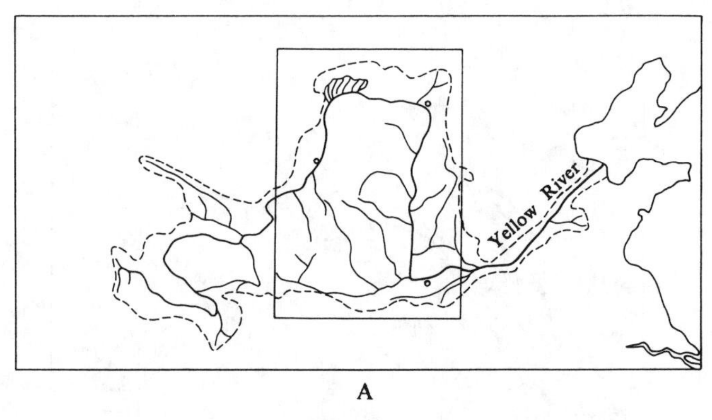

A

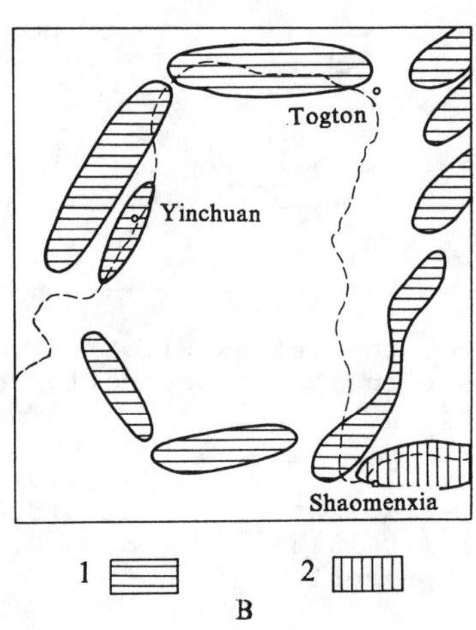

B

Fig.5 Evolution of the water system in midstream of the Yellow River Delta during Quaternary.
  A. Unit Yellow River system at present.
  B. Seperate pre-water-system of Yellow Riveer in Pleistocene.
  1. Lake basin around the margin of the Ordos subplate.
  2. Shaomenxia lake basin near the boundry of North China Plain.

changes to the south and flows into the Yellow Sea for 728 years. In 1855 the river burst its bank near Zhengzhou and turned to the north again (History outline of Yellow River water conservancy editorial commission,1982, Fig.6).The Flooding water of Yellow

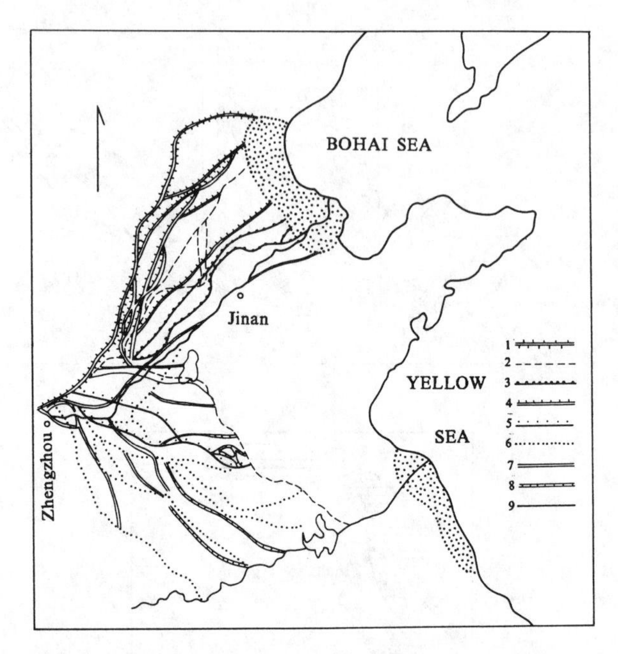

Fig.6 Migration of the Yellow River in the last 3000 years and the distribution of the yellow River Delta complex.

1. Before 602 B.C.
2. 602 B.C.-A.D.11
3. A.D.11-A.D.1048
4. A.D.1048-A.D.1127
5. A.D.1127-A.D.1234
6. A.D.1234-A.D.1391
7. A.D.1391-A.D.1546
8. A.D.1546-A.D.1855
9. A.D.1855-present

River fell to the existing Daqing River that was three hundred meters wide and six to nine meters deep and emptied into the Bohai Sea. No mouth bar was there in the estuary. As the Yellow River fell in, large amounts of sediments deposited in the channel and the estuary. The channel became shallow and narrow and shifted in 1889. The life of the first ditributary is 34 years. This is the longest life then the later other nine distributaries.

With the migration of distributary channel the deltaic lobes grow up one by one. Six lobes were formed in 855-1934 with its apex of Ninghai and four lobes were formed from 1934 to present with a apex of Yuwa (Cheng 1991, Fig.7). The average life of a lobe is only

13 years.

The evolution of a deltaic lobe can be divided four stages. In the first stage the lobe begins form and the deltaic plain expands seawards rapidly, when the channel is braided shifting and widening. In the second stage the river mouth bar grows up with high accumulation rate, while the braided channel converge to a straight one. In the third stage, the mouth bar lateral expands as the channel curve and burst to the both sides of the mouth bar. The four stage is erosion stage while the river shift out of this lobe. Finally sand ridge is formed at the coastline.

The coastline prograde seawards several kilometers per year at first and second stages and retreat landwards hundred or tens meters at fourth stage(Fig.8)

The sedimental environments of an active lobe includes five division shown in Fig.9. Compared to other delta the lateral-delta depositional environment, the mud deposition unit, is a special division.
It is formed on both sides of the mouth bar, because of

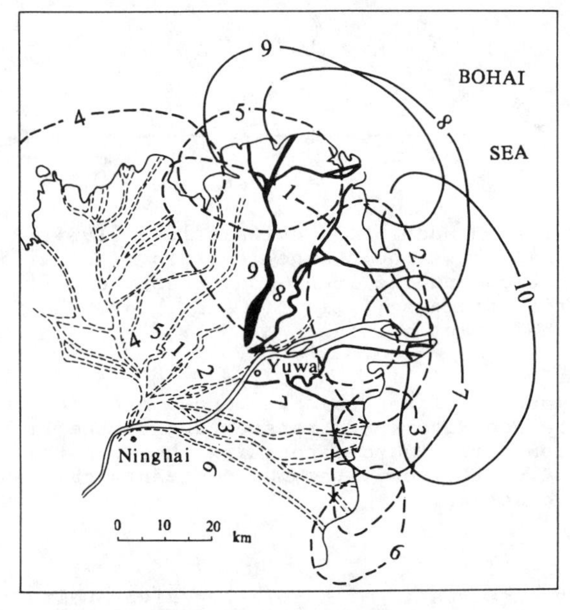

Fig.7 Shifting of distributary channel and deltaic lobes of the modern Yellow River Delta. The active period of channels and lobes is as follows:
(1) 1855.6-1889.3 (2) 1889.3-1897.5 (3) 1897.5-1904.6
(4) 1904.6-1626.6 (5) 1926.6-1929.8 (6) 1929.8-1934.8
(7) 1934.8-1953.7 (8) 1953.7-1964.1 (9) 1964.1-1976.5
(10) 1976.5-present (from Cheng 1991)

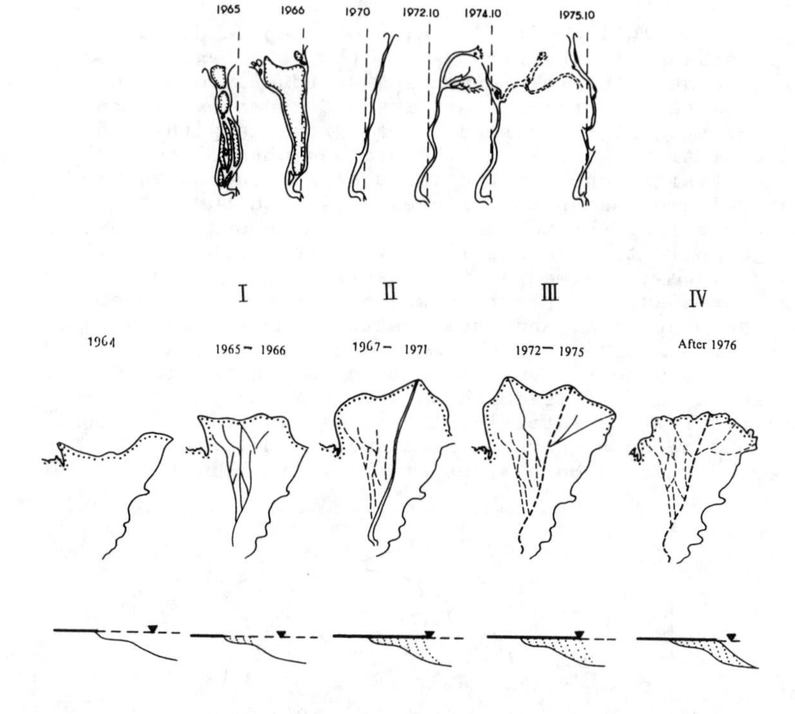

Fig.8 Evolution Model of a deltaic lobe (1964-1976). The upper map shows the channel shifting, the middle is coast migration,and the lower indicates the sedimentation process I, II, III and IV are the four stage of the evolution.

a large amount of fine grain suspended material coming from the Yellow River and transporting by the tidal wcirculation current and depositing there. After abandonment of the lobe, the mud sediments carried out by the wave action.

CONCLUSIONS

The characteristic of the Yellow River Delta is frequently migrating of the channel and the distributary, the coastline as well. The major reason of this feature is high suspended load of the river and extremely shallow of the Bohai Sea. The average annual sediment discharge is $1.18x10^9$ tons and the annual

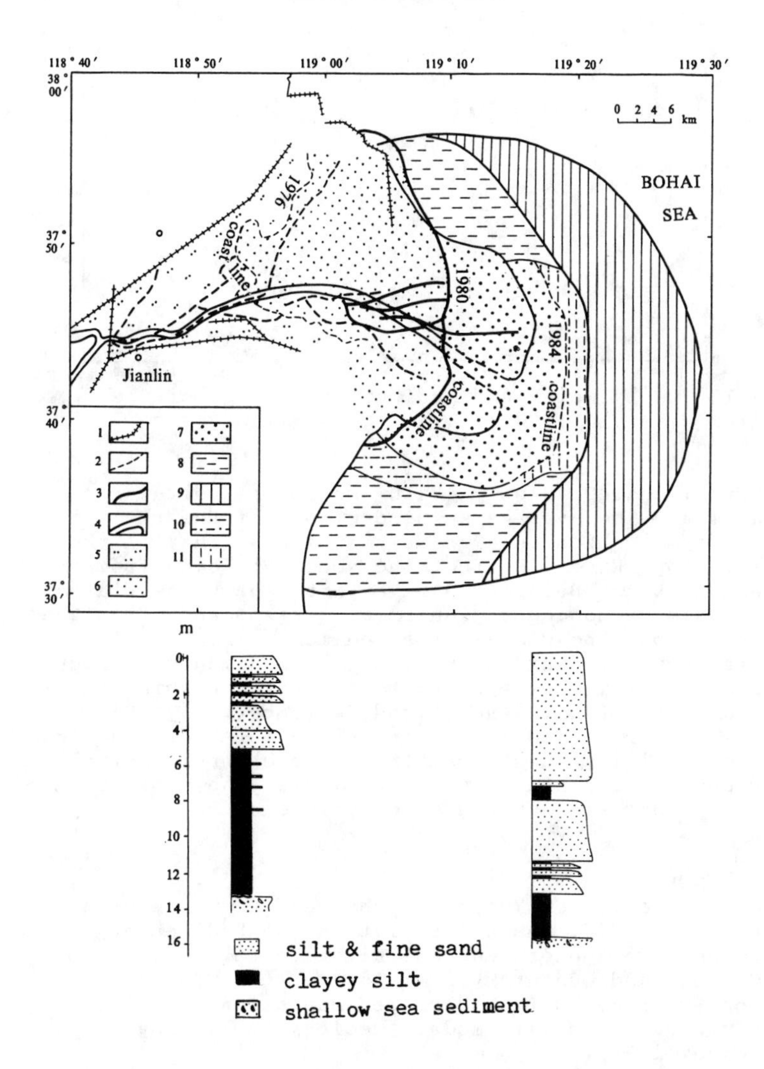

Fig.9 Sedimental environments and vertical sequence of the active lobe.
1. Artificial bank 2. Ancient channel(1976-1979)
3. Ancient channel(1980-1984) 4 active channel
5. fluvial division 6. Mouth sheet division
7. Mouth bar division 8. Lateral-delta division
9. Prodelta division 10.Transitional zone
11.Distal bar division

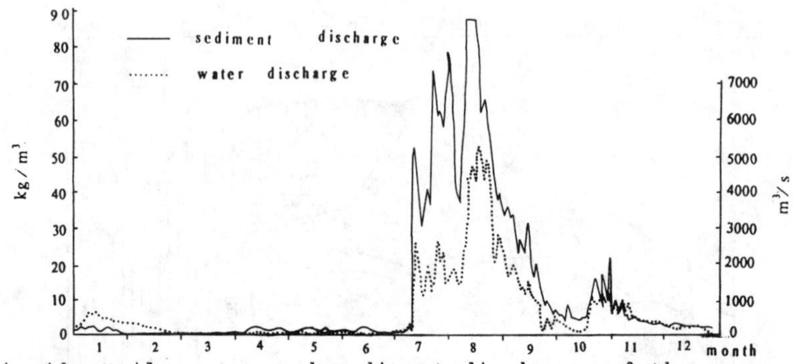

Fig.10 Daily water and sediment discharge of the Yellow River in 1988.

water discharge is only 42x10^{9} m3 in the Delta area. That means the average suspended load of the river is 25.3 kg/m3. which is 42 times as high as that of the Mississippi River. Most of the sediments come from the Loess Plateau. Before the river into the Loess Plateau the sediment concentration is 5.73 kg/m3 and after the river across the Plateau the sediment concentration increase to 25.3 kg/m3. Most of the sediments , about 90% of total,is transported to the Delta within the flood season (July, August and September, Fig 10). A large number of sediments accumulate in a very shallow basin resulting the migration of the channel. Actually, the channel migration is a function as the erosion rate on the Loess Plateau and space volume on the estuary.

REFERENCE

Cheng Guodong, Ren Yucan, Li shaquen, Li Guangxue and Dong Wan,1986. Channel Evolution and Sedimentary Sequence of Modern Huanghe River Delta. Marine Geology and Quaterary,2: 1-15.

Cheng Guodong, 1991. Modern Sedimentation of Yellow River Delta and its Model. Geology Publishing House. 109pp.

History Outline of Yellow River Water Conservancy Editorial Commission, 1982. History Outline of Yellow River Water Conservancy. Water Conservancy and Electric Power Press. 397pp.

Li Desheng, 1981. Geological Structure and Hydrocarbon Occurrnce of Bohai Gulf Oil and Gas Basin. Marine Geological Research. 1: 3-20.

Natural Geography of China Editorial Commission, 1982. Natural Geography of China (Historical Geography of

China). Science Press. 28-86pp
Pan Guitang, Wang Peisheng, Xu Yaorong, Jiao Shupei and Xiang Tianxiu, 1990. Cenozoic Tectonic Evolution of Qinghai-Xizang Plateau. Geological Publishing House. 190pp.
Qin Yunshan, Zhao Yiyang, Chen Lirong and Zhao Songling, 1990. Geology of Bohai Sea. China Ocean Press. 232pp.
Yang Zigeng and Lin Hemao, 1991. Quaternary Processes in Eastern China and their International Correlation. Geological Publishing House. 89-115pp.
Zhang Kang,1989.Tectonic and Resources of Ordos Fault-block. Shanxi Science & Technic Press. 289-304pp.

Relative Sea Level Rise and Coastal Erosion in the Yellow River Delta, China and Management Strategies

Ren Mei-e[1]

Abstract

The Yellow River Delta consists of 3 deltas: the Ancient, Modern and abandoned Delta. In the Ancient Delta, serious land subsidence due to ground water overpumping resulted in relative sea level rise (at Tanggu) of 26 mm/yr in the last 30 years and coastal hazard is aggravated. In the Modern Delta, frequent shift of lower Yellow River course (about once every 12 years) renders the area extremely unsafe for development. In the Abandoned Delta, owing to sediment starvation, the coast has retreated about 20 km since 1855 and 1400 sq. km of land were lost.

Response strategies include strict control of ground water pumping, stabilization of the present lower Yellow River Channel and improvement of coastal protection works.

General Geographical Setting

The Yellow River Delta, China is a unique delta in the world (Grabau, 1936). Owing to its huge sediment load ($16x10^8$ tons at Sanmenxia) and frequent shift of its lower course over an extensive area spanning more than 6 degrees latitude, the Yellow River has built a huge Holocene delta

---

[1] Professor, Department of Geo and Ocean Sciences and State Pilot Laboratory of Coastal and Island Exploitation, Nanjing University, Nanjing 210008, China.

with an area of 250,000 km^2. In the coastal zone, 3 recent deltas may be distinguished; from north to south, they are (Fig. 1, 2):

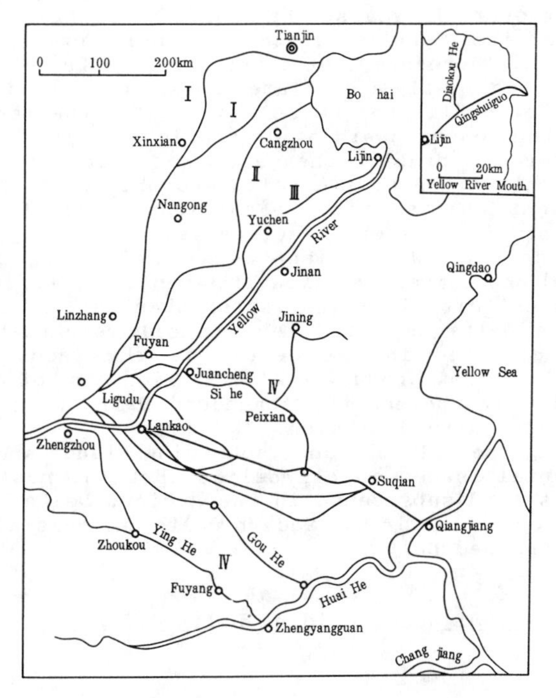

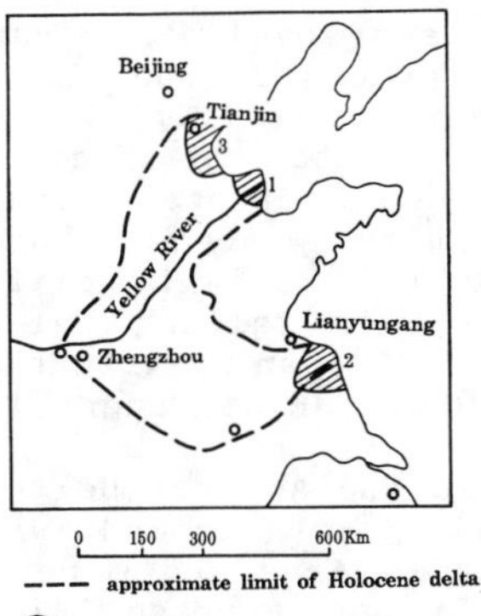

(1) Ancient delta in the vicinity of Tianjin, formed between 3400 B.C. and 1128 A.D. Area about 8,000 km^2.

(2) Modern delta, formed since 1855 A.D. Area about 5,400 km^2.

(3) Abandoned delta, formed between 1128 and 1855 A.D. Area about 7,100 km^2.

During the last 5,000 years, 26 changes of major channel of the lower Yellow River had occurred, the most important was the shift in 1128-1855 when its course was directed southward from Ligudu, Henan Province to the Yellow Sea.

In the last 140 years (1855-present), the lower Yellow River in the Modern Delta, owing to small flow and huge sediment load[2], has shifted its course 11 times, averaging about once every 12 years. The unsafe environment rendered this otherwise fertile delta extremely precarious for development. Therefore, until the early 1960s', the Modern Delta was practically a desolate, swampy wasteland. Changes of the river outlet to the sea especially the last change from Diaokou He southward to Qingshuigou in 1976 (Fig. 1), has triggered a drastic change of sediment budget in the coastal zone of the Modern Delta and the southern Part of the Ancient Delta with significant effect on the development of ports on these coasts.

In economy, the Ancient Delta is a well developed industrial and commercial area. Its center, Tianjin, is the third largest city in China (after Shanghai and Beijing). The Modern Delta is noted for its oil resources.Shengli Oilfield, located in the delta, is the second largest oilfield in China, producing 33.5 million tons of crude oil a year (1991). Whereas the Abandoned Delta is a populous and rich agricultural region.

Rising sea level and subsiding land are common environmental issues in all deltas. But the most serious problem is land subsidence in the Ancient Delta, shifting Yellow River course in the Modern Delta and coastal erosion in the Abandoned Delta.

Rising Sea Level and Subsiding Deltas

In the last 30 year,relative sea level at Tanggu in the Ancient Delta is rising but stations on the coast of Shandong peninsula have falling sea levels (Fig. 3, Table 1)(Ren, 1993). This is chiefly due to the fact that deltas are located in areas of recent crustal sinking, rate of sinking generally 2 mm/yr whereas Shandong Peninsula is a crystalline craton with slight uplifting generally 2 mm/yr. Other stations with falling relative sea level include Qinhuangdao, located in an uplifting area of the Yanshan Mountains, about 200 km NE of Tanggu(Ting, 1992).

Since 1959, land subsidence in the Ancient Delta was greatly accelerated by overpumping of ground water, reaching 81.6 mm/yr in Tianjin Urban Area and more than 100 mm/yr in the coastal town of Tanggu (average 1970-

---

[2] At Lijin,near the apex of the Modern Delta and about 100 km from the sea, annual flow of the Yellow River 427×10^8 m^3, suspended sediment load 10.6×10^8 tons (average 1949-1985) about 8% and 500% that of the Mississippi respectively.

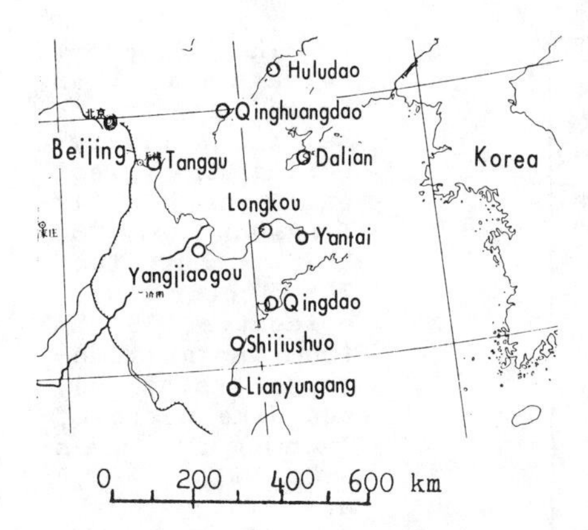

Fig. 3 Tide-gauge stations in North China

1984). It has been calculated that 88% of the total land subsidence in Tianjin New Port Area is due to ground water pumping (the other 12% due to tectonic subsidence and sediment compaction).

Rate of land subsidence varies greatly both in time and space. In 1986, owing to water supply from the Luan He, land subsidence at Tanggu tide-gauge station was reduced to 4 mm but in a greater part of Tianjin New Port area, it still amounted to about 20 mm. In Hangu and Dagang, a little north and south of Tanggu, land subsidence in 1986 remained at about 100 mm/yr(Fig. 4, 5).

Table 1 Relative sealevel changes in selected tide-gauge stations in Yellow River Delta and its adjacent area.

No.	Station	Period	Rise(+)or Fall(-)	Rate (mm/a)
1	Qinhuangdao	1960-1989	-	1.95
2	Tanggu			
2(1)	Beipotai	1910-1936	+	1.77
2(2)	Liumi(6 m)	1959-1985	+	1.20
3	Yangjiaogou	1952-1986	-	1.20
4	Longkou	1961-1989	-	1.31
5	Yantai	1960-1989	-	3.39
6	Qingdao	1952-1985	-	1.07
7	Shijiusuo	1968-1989	-	0.85
8	Lianyungang	1963-1985	-	1.86

Note: rate of sealevel changes from results of linear regression analysis, data of Tanggu (Liumi) Station 1959-1985 have been corrected for land subsidence and are therefore not relative sea level.

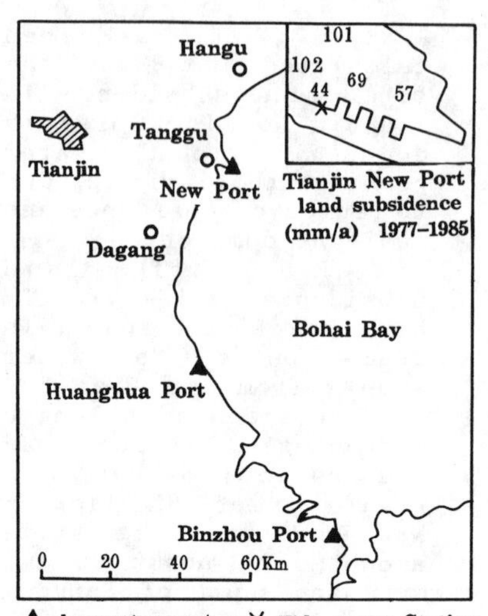

Fig. 4 Coastal region of Tianjin

The Ancient Delta is also an area of strong seismic activity. During the great earthquake at Tangshan, July 28, 1976 (Magnitude 7.8 Richter scale, epicentre 78 km from Tanggu tide-gauge station), sea dike north of Tanggu was subsided 1.3-1.5 m.

From the available data, the following estimate of relative sea level rise in the last 30 years and next 20 years is made (as shown in Table 2).

In table 2, estimates of the past and future eustatic sea level rise are the lowest figures (Stewart et al, 1990; IGBP, 1992). Past land subsidence rate in the Ancient Delta after the result of precise levelling from benchmark at Tanggu tide-gauge station to the master benchmark on a rocky hillock (stable benchmark) in Baodi north of Tianjin City, between 1966 and 1985 (levelling every year). As land subsidence in the Ancient Delta is mainly man-induced, it is difficult to make any accurate estimate for the next century, because the amount of ground water withdrawal in the future depends largely on a number of socio-economic factors and on the policy of the local authority. The present estimate is based on the consideration that the area is now seriously short in water resources; although water transfer from the Yangtze River will considerably increase water supply in the future, rapid growth in population and economy (industrial production is likely to grow at a rate of 10% per year) in the next 20 years will greatly augment water demand. It is expected that ground water pumping will be considerably

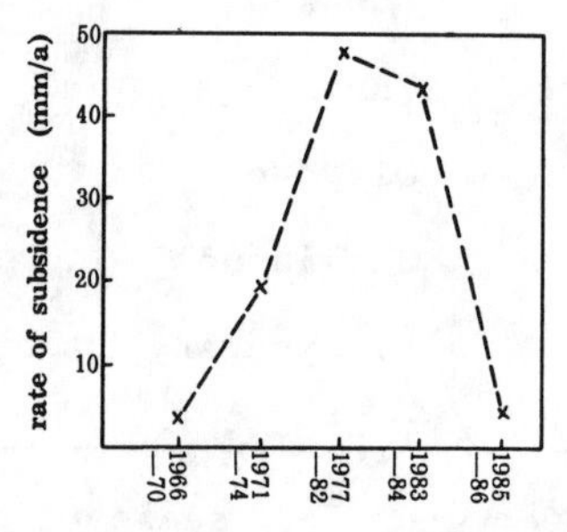

Fig.5 Rate of subsidence of benchmark at Tanggu tide-gauge station, 1966–1986.

reduced but is not likely to cease. Land subsidence 10 mm/yr is probably a best estimate.

Table 2 Estimate of relative sea level rise (mm/yr) in the Yellow River Deltas.

Area	eustatic sea level rise	land subsidence	relative sea level rise
A. The last 30 years			
Ancient Delta	1.5	24.5	26.0
Modern Delta	1.5	2.0	3.5
Abandoned Delta	1.5	2.0	3.5
B. The next 20 years			
Ancient Delta	5.0	10.0	15.0
Modern Delta	5.0	2.0	7.0
Abandoned Delta	5.0	2.0	7.0

There are no tide-gauge stations with long records in the Modern and Abandoned Delta. Rate of land subsidence in these deltas is from Geodetic Survey, State Seismological Bureau of China, 1951-1982 (Ting, 1982).

Storm surge is an important factor in the evaluation of risks in the future sealevel rise. The Ancient Delta is an area of severe storm surge. According to ancient Chinese historical document, Chronicle of the Han Dynasty (written in 1st century A.D.), the whole Ancient Delta below the present 4 m contour was inundated by the sea during a disastrous storm surge in about 2000 BP. In a recent strong storm surge, August 11, 1938, the sea also advanced to the 4 m contour near Huanghua City. In view of large rate of relative sea level rise, the whole Ancient Delta below the 4 m contour (area 7,000 sq. km) is vulnerable to sea level rise. About 9 million people, 12 billion US dollar annual industrial production (1990) including large power plants, petro-chemical and chemical factories and a coastal oilfield producing 4 million tons of oil a year will be affected.

The Modern Delta is also affected by severe storm surges. During strong storm surge, the water level was raised 277 cm near Binzhou Port in April 5, 1964 and 373 cm at Yangjiaogou in April 23, 1969. Therefore, area below the 3 m contour (about 3000 sq. km) should be regarded as vulnerable. Although population affected will be small (about 1 million), large coastal oilfields producing 15 million tons of crude oil a year will be in danger (Ren and Cui, 1991).

Coastal Erosion and Port Development

Coastal erosion in the Yellow River Delta is mainly caused by sediment starvation rather than by sea level rise. The most remarkable case is the Abandoned Delta. As the lower Yellow River was shifted northward to the Bohai Sea in 1855, fluvial sediments of nearly 1 billion tons/yr were no longer being delivered to the coastal waters of the Abandoned Delta. As a result, the coast changed immediately from progradation to retrogradation. In the early years of the river abandonment, rate of coastal recession reached approximately 1000 m/yr, the highest rate of coastal erosion documented in China. In the last 20 years, erosion has affected 188 km delta coast but rate of coastal recession has decreased to about 20 m/yr. Altogether, the coast of the Abandoned Delta has retreated about 20 km since 1855 and approximately 1400 sq. km of land were lost (Ren, 1992).

Shift of the lower course of the Yellow River in the Modern Delta has profound effect on littoral sediment drift and coastal erosion. With the shift of the river course from the Diaokou He to the Qingshuigou in 1976, most of the river flow and sediment are diverted toward south. This situation is accentuated by unique oceanographic conditions in the coastal waters: an amphidromic point with associated swift current field (maximum current speed 180 cm/s) at 38°08'41"N, 119°08'57" (near Huanghe Sea Port) (minimum tidal range 0.8 cm). The swift current field forms a cell around which residual current flow in a clockwise direction. On the south side of the cell, clear sea water flows landward, acting as a hydrographic barrier against the northward diffusion of the fresh water and silt from the Yellow River. As a result, a stationary turbidity front is formed near 37°50'(Ren and Shi, 1986). Investigation of 166 NOAA imageries between June, 1983 and December 1984 verifies this conclusion (Jiang and Wen, 1987).

The coastal environment of the Modern Delta changes

rapidly with change of the position of the outlet of the Yellow River. With the shift of the river course to the Qingshuigou in 1976, the northern coast of the Modern Delta together with its adjacent coast of the Ancient Delta have change from progradation to retrogradation. According to data of satellite imageries, the overall rate of coastal recession between August, 1977 and September, 1984 was 345 m/yr for 44 km of the coast near Huanghe Sea Port and 139 m/yr for 38 km of the coast east of Binzhou Port where the Yellow River abandoned its course in 1926 (Chen, 1989).

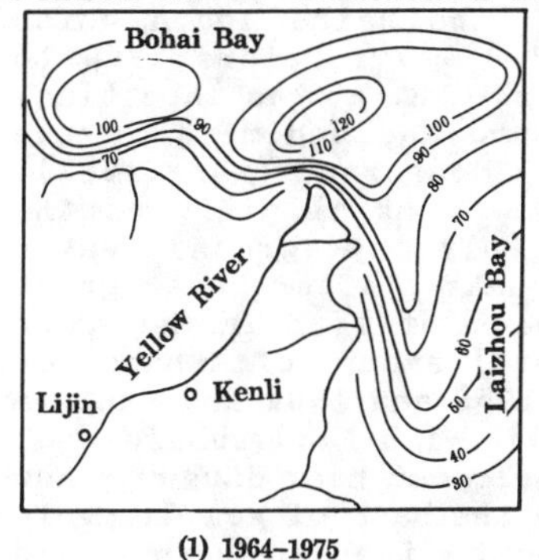

(1) 1964–1975

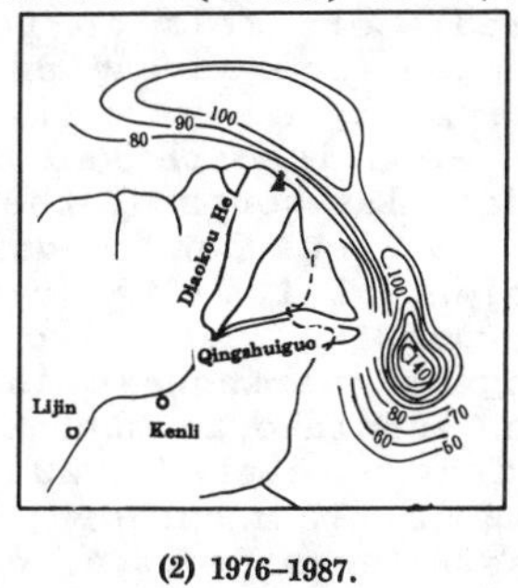

(2) 1976–1987.

Huanghe Port

Fig.6 Current along the coast of Yellow River Delta (in cm/s)

Near the mouth of the Diaokou He, immediately after the river abandonment (in 1976), the coast was retreated 5.0-5.5 km between 1976 and 1984 and the adjacent sea bottom between 0 and -13 m was scoured and lowered at a rate up to 50 cm/yr (Geng, 1988). Further west, the coast near Binzhou Port and Huanghua Port is also retrograding. Low cheniers were formed near the high water and sediments of the tidal flat are coarsened. On the coast near Huanghua Port, erosion bluffs (0.2-1 m high) are formed and watching towers built in 1967 on the cheniers near the sea were collapsed due to coastal erosion. At the same time, the adjacent shallow sea bottom is scoured and isobaths are retreating landward. Comparison of bathymetric charts between 1959 and 1984 indicates that the -2 m isobath was retreated 14.96 m/yr and 2 m/yr and -4 m isobath retreated 9.40 m/yr and 2 m/yr near Binzhou Port and Huanghua Port respectively.

However, since 1986, the construction of strong dikes along a great part of the coast between the present Yellow River mouth and Binzhou port has greatly checked coastal erosion and the rate of coastal recession is reduced to about 1-2 m/yr.

A part of the coast of Tianjin has also witnessed a change from accretion to erosion in the last 30 years, due chiefly to the reduction of riverine sediment supply and large land subsidence. Since the 50s', the construction of reservoirs in the upper and middle reaches of the river and sluices and locks at or near the river mouths have deprived a major part of fluvial sediments from entering the sea. For example, before 1958 when reservoirs and lock were built, the Hai He, the largest river in Tianjin area, delivered 813×10^4 tons of sediment to the sea every year (average 1917-1958) near Tanggu; after 1958, its annual sediment input was reduced to only 27.9×10^4 tons (average 1959--1987). The effect of sediment starvation was immediately felt in Dagu Bank at the Haihe mouth which changed from advancing seaward 77.8 m/yr before 1958 to retreating landward 69.5 m/yr after 1958. At the same time, the -2 m isobath was moved landward and sediment of the tidal flat south of the Haihe mouth coarsened (Wang,1987).

Serious land subsidence in the coastal plain of the Ancient Delta is caused not only by ground water overpumping but also by earthquake. After the great Tangshan earthquake in 1976, a part of Tianjin New Port area subsided as much as 1.2 m. Total amount of subsidence in Tanggu reached 2.298 m between 1959 and 1984 and that in Hangu 2.137 m between 1957 and 1983. With land subsidence, coastal erosion is accelerated. The coast near Hangu is now suffering from serious erosion and the base of sea dikes is badly scoured. In the offshore, the 0 m isobath was receded 400-1400 m between 1958 and 1983.

Sediment starvation and coastal erosion have a significant effect on port construction and development. Tianjin New Port (handled 24 million tons of cargo, 1991), one of the leading ports in China, is an artificial harbour built on tidal mud flat. Since its construction in 1939, it was troubled by serious siltation problem. Sediment source of the harbour siltation came chiefly from the Haihe which enters the Bohai just south of the port. Since the building of Haihe lock in 1958, nearly 99% of Haihe sediments were cut off and consequently, amount of siltation in the harbour was reduced by 1/3. After successful expansion and improvement in the last 40 years, Tianjin New Port is now accessible to ships of 50,000 tons. British luxury liner Queen Elizabeth II (67,100 tons) called at the port's container berth in March, 1988. Detail study demonstrated that in the present sediments responsible for harbour siltation come from silts on the offshore sea bottom resuspended by waves. Therefore, it is planned to extend the breakwaters to beyond the breaker zone where suspended sediment concentration is lower so that siltation in the harbour could be further reduced.

The success of Tianjin New Port provides a convincing example that deep-water ports can be built on the mud plain coast and successfully managed, provided there is no large littoral sediment drift (Wang, 1987).

Sediment starvation and coastal erosion also encourage

port development in other parts of the Yellow River Delta. Huanghua Port, about 70 km south of Tianjin New Port, is currently expanded to accommodate ships of 50,000 tons. Huanghua Sea Port (38°05', 118°57'), about 45 km north of the present Yellow River mouth, has had practically no siltation since its construction in 1984. It is now accessible to ships of 3000 tons and a sea ferry service between the port and Lushun (Liaoning Province) on the other side of the Bohai was opened in August, 1992. In the Abandoned Yellow River Delta, owing to continuous serious coastal erosion since 1855, coastal water is very deep. On the 35 km coast south of the Abandoned Yellow River mouth, the -10 m isobath is only 3.5 km from the coastline in certain most favourable sections. Plan to build a deep water port here is being considered.

Management Strategies

Dynamic environment of the Yellow River Delta presents favourable opportunities as well as serious challenges. To minimized the adverse effect of the latter, the following management strategies are suggested:

(1) Shifting lower Yellow River is extremely unfavourable to sustainable development of the Delta. Since 1985, with rapid development of oil fields and petrochemical industry, it is necessary to control and stabilize the lower course of the Yellow River in the Modern Delta. Various engineering measures were taken: extending river dikes to the estuary, straightening meandering river course, damming minor tributaries and branches, deepening river mouth bar by dredging etc. The result is encouraging As the river flow is confined to a single, straight channel, its flushing power is increased, more sediments are being carried to the sea and the channel becomes relatively stabilized. Detailed investigation by Yellow River Water Conservancy Commission of China in December, 1991 (with use of hydrologic models) demonstrates that through engineering measures, the present lower Yellow River course in the Qingshuigou may be stabilized for another 30 to 50 years. It is urged that more attention should be paid and sufficient fund allocated to the control of the lower Yellow River Channel.

As the stability of the lower Yellow River course depends largely on its sediment load, it may be noted that owing to the construction of large hydro-power stations and thermal power plants, development of coalfields and expansion of irrigation works in the river basin, sediment load of the river was decreasing in the last decade[3]. This

[3] Annual suspended sediment load at Lijin 7.75×10^8 tons in 1971-1987 against 12.74×10^8 tons in 1950-1970.

will be favourable to the stability of the river channel in the future.

(2) Owing to serious land subsidence, elevation of over 50% of Tianjin area is now less than 1-3 m a.s.l. and the lowest part, Riverside Park in Tanggu town, is 3.3 below sea level. Thus, coastal hazard from sea level rise and storm surge will be aggravated.

On Tianjin coastal plain, because of continuous overpumping of ground water, three formerly separated ground water funnels have now become united, covering an area of 1270 sq. km . Its water table in 1985 was 67-77 m below sea level. Evidently, this will increase hazards from salt water intrusion. Thus, in the Ancient Delta, strict measures should be taken to limit ground water withdrawal. The existing technique of water reuse should be more widely practised. At the same time, the work for an alternative source of water supply, water diversion from the Yangtze River, should be accelerated.

(3) While coastal erosion on the Yellow River Delta may be helpful to port development, it is a serious threat to the community at large. It is important that solid coastal defense works should be constructed to protected the weakly protected and unprotected eroding coast.

Concluding Remarks

Frank Press, the president of U.S. National Academy of Sciences, has pointed out that "humankind has become a more important agent of environmental change than nature" (Press, 1989). The Yellow River Delta provides an excellent example. The shift of the lower Yellow River course from the Bohai to the Yellow Sea in 1128 A.D. was triggered by human breaching of dikes for military reasons. The shift of the river channel in the Modern Delta in 1976 was also a human diversion. Its effect on coastal accretion and erosion has important socio-economic consequences. In Tianjin coastal plain, man's unwise use of ground water (overpumping) has caused large land subsidence and coastal hazards are aggravated. On the other hand, the successful stabilization of the lower river course in the Modern Delta since 1988 demonstrates what man can do to control the shifting river course which was formerly regarded as a "law of nature".

The past experiences and lessons may prove valuable to the wise management of the Yellow River Delta.

References

Chen, Shupeng, (ed.), 1989, Papers on the Application of Land Resources Satellite Data, vol. 2, pp. 40-43. Science Press, Beijing.

Geng, Xiushan and Wu, Shiying, 1988, Preliminary approach to the dynamic geomorphological conditions of Huanghe Sea Port, Acta Geographica Sinica, 43(4):299-310.

Grabau, A.W., 1936, The Great Huangho Plain of China, Journal of the Association of Chinese and American Engineers, Peking, quoted in Christopoulos, C.P., 1947, Great Plain-Building in North China, Bull. Geol. Soc. China, 27, p.305.

IGBP, 1982, Global Change: reducing Uncertainties, the International Geosphere-Biosphere Programme, the Royal Swedish Academy of Sciences, Stockholm, p.21.

Jiang, Zhongxi and Wen, Linping, 1987, Analysis of the effects of Suspended Sediment of Huanghe River on harbour construction by using NOAA images, Oceanologica et Limnologica, 18(3), 244-248.

Li, Z.G., 1991, Current field in the offshore of the Yellow River mouth, Yellow River Delta Research. May, 1991,15-20, Dongying, Shandong Province.

Press, F., 1989, What I would advise a head of state about global change, Earth Quest, 3(2):1-2.

Ren, Mei-e, 1992, Human impact on coastal landform and sedimentation----the Yellow River Example, GeoJournal, 28, in press.

Ren, Mei-e, 1993, Relative sea level changes in China over the last 80 years, Journal of Coastal Research, in press.

Ren, Mei-e and Cui,Gonghao, 1991, Relative sea level rise in Yellow River Delta--implications and response strategies, Journal of Chinese Geography, 2(2):1-13, Singapore.

Ren, Mei-e and Shi Yunliang, 1986. Sediment discharge of the Yellow River and its effect on the sedimentation of the Bohai and Yellow Sea, Continental Shelf Research, 6, 785-810.

Stewart,R.W., Kjerfve,B., Milliman,J. and Dwivedi,S.N., 1990, Relative sea level change: a critical evaluation, UNESCO report in marine science, UNESCO.

Ting, G.Y. (ed.), 1991, A Treatise on geodynamics of the lithosphere of China, pp.123-141, Seismological Press, Beijing.

Wang, Rukai, 1987, The review and prospect of siltation on Tianjin Port, in Proceedings of Coastal and Port Engineering in Developing Countries, Vol.2, 1478-1489, China Ocean Press, Beijing.

Evolution of the Santee/Pee Dee Delta Complex, South Carolina, USA

Miles O. Hayes[1], Walter J. Sexton[2], Donald J. Colquhoun[3], and Timothy L. Eckard[3]

Abstract

The Santee/Pee Dee delta is located on the north-central portion of the Lower Coastal Plain physiographic province of South Carolina, USA, which is underlain by Pleistocene strand plain and deltaic sediments. During the Late Wisconsin and earlier sea-level lowstands, major river valleys were incised into the underlying sediments to depths approaching 30 meters (m), responding to base level changes of near 100 m between interglacial and glacial times. Basal fluvial sands grading upwards into deltaic sands and muds filled the drowned river valleys as they were flooded during the Holocene sea-level rise.

Morphologically, the delta is classified as a "mixed-energy" delta, because wave and tidal forces mold the outer margin of the delta with equal efficiency, and because the sediment load of the two rivers that build the delta, though by far the largest of any of the river systems of the Georgia Bight, is modest compared to other rivers on a global scale. The upper delta plain is composed of a coarse-grained sand meander belt confined within the incised alluvial valley. The lower delta plain sediments are dominated by mud, and extensive marshes cover the fan-shaped delta surface. The delta front is a huge, arcuate, sand-dominant bulge in the shoreline intersected by three large tidal inlets, which have massive ebb-tidal deltas containing hundreds of thousands of m^3 of sand. Stunted delta-front barrier islands flank the inlets.

Stratigraphically, the delta is made up of two distinct components, the delta plain (upper and lower) and the delta front deposits. The

----

[1]Research Planning, Inc., 1200 Park Street, Columbia, SC 29201
[2]Athena Technologies, Inc., 3700 Rosewood Drive, Columbia, SC 29205
[3]University of South Carolina, Columbia, SC 29208

Holocene sediments of the delta plain are mostly alluvial valley fill sediments composed of multistory and multistage point-bar and channel sands. These coarse-grained sediments are overlain by marsh and tidal flat mud deposited as a result of the elevated sea level. The Holocene stratigraphy of the delta front is exceedingly complex, being composed of tidal-inlet fill, barrier island, and aggraded alluvial valley deposits.

Since the beginning of historic times, the upper surface and river flow of the delta have been modified dramatically by man. From the late 1600s to the late 1800s, extensive rice farming was carried out on the delta, practices which modified the natural sediment and vegetation, with the most notable effect being the deforestation of an extensive cypress/tupelo swamp. Construction of hydroelectric dams upriver in the early 1900s steadily reduced the sediment supply to the delta, culminating in 1939 with construction of a large dam which diverted 90 percent of the original discharge of the Santee River into the Cooper River and ultimately into Charleston Harbor. The results of the diversion included: 1) significant erosion of the delta front; 2) increased salinity upriver, with associated changes in biological communities; and 3) increased siltation in Charleston Harbor. In 1986, a variable percentage of the flow of the Santee River above the dam was diverted back to the river via a newly constructed rediversion canal. Whereas the outer margin of the delta is a wildlife refuge and presently subject to little human modification, the man-induced changes in river-flow volumes have brought about some remarkable changes in the morphology, stratigraphy, and ecology of the delta.

Introduction

The Santee/Pee Dee delta is located on the north flank of the Georgia Bight (Figure 1). River deltas are relatively scarce in the Bight, because in order for a prograding delta form to have been developed on the present shoreline of the Bight, the alluvial valley eroded during the Wisconsin and earlier lowstands had to be aggraded to above sea level beyond the present strandline. Where this has not happened, estuaries occupy the drowned river valleys. The formation of such estuaries is primarily the result of insufficient sediment load of the rivers. However, the process may be aided by the general tectonic downwarp in the region and by the strong tidal flow that transports the river sediments offshore, which is probably the situation in some of the large estuaries at the head of the Bight, where spring tidal ranges exceed 2.5 m (Figure 1).

Only three river systems have built what would be referred to normally as ocean-front deltas in the Georgia Bight: the combined Santee and Pee Dee Rivers in South Carolina; the Savannah River at the South Carolina/Georgia border; and the Altamaha River in Georgia. These

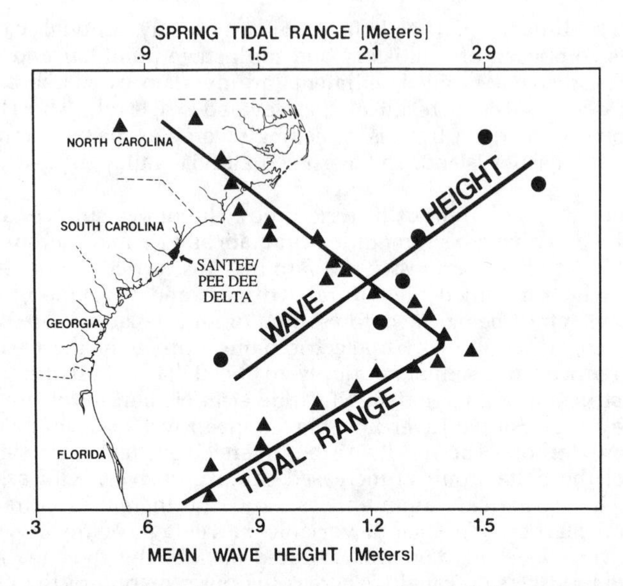

Figure 1. Location of the Santee/Pee Dee delta, and the variation of wave height and tidal range along the shoreline of the Georgia Bight.

three piedmont river systems have the largest sediment loads of the rivers within the Bight. The Santee/Pee Dee River system has formed by far the largest delta of the three, but this delta is only small- to medium-sized in comparison with other deltas of the world.

Classification

Most present-day workers, following the original ideas of Price (1955) and Bernard (1965), focus on three basic factors—sediment supply, wave energy, and tidal-current energy—in defining the morphostratigraphic character of deltas. Galloway (1975) placed this concept on a ternary diagram, with the three basic factors occupying the three poles of the diagram. The Santee/Pee Dee delta is classified as a mixed-energy delta, as it plots halfway between the tidal- and wave-energy poles and toward the "destructive" side of Galloway's diagram. As illustrated in Figure 1, the delta is located on the portion of the Georgia Bight that has a near-mesotidal range and moderate waves; consequently, marine processes mold the outer margin of the delta. The sediment load of the two river systems (about 5 x 10 tons of suspended sediment influx/yr before modification), though the largest in the Georgia Bight, is modest compared to other rivers on a global scale.

Delta-Plain Morphology

The upper delta plain, the landward limit which coincides roughly with the landwardmost influence of the tides, is composed of a coarse-grained meander belt confined within the alluvial valley. Fluvially dominant sedimentary environments are present, including channel (point bar and oxbow), channel bank (levee), and floodplain (backswamp) deposits. These depositional environments are characterized by abrupt lateral changes in sediment types.

The lower delta plain is a 10-kilometer wide triangular belt of mud-dominant environments. Today, the surface of the lower delta plain consists of a checkerboard pattern of refurbished rice fields encased in brackish and saltwater marshes. Before rice cultivation, a freshwater swamp extended to near the present North Santee River Inlet (Figure 2).

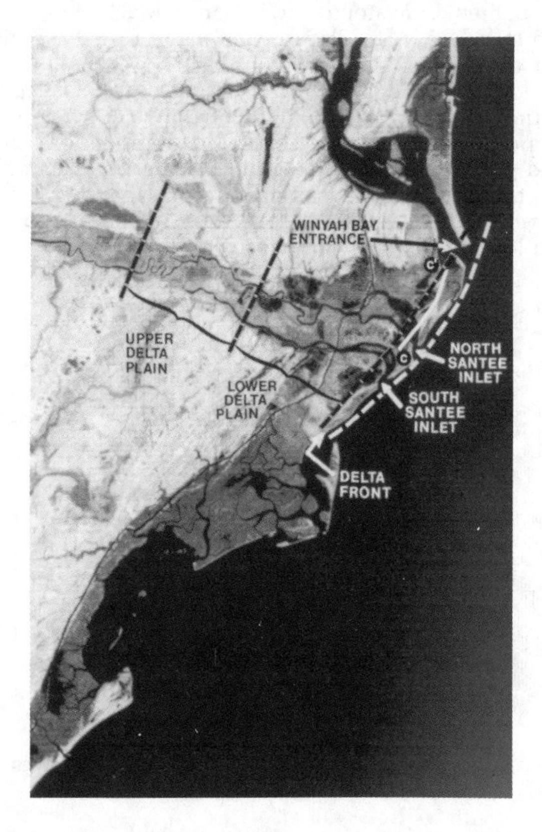

Figure 2. Satellite infrared image of the Santee/Pee Dee delta region.

Delta-Plain Stratigraphy

The nature of the stratigraphy of the Santee/Pee Dee delta is complex, but it has been deciphered on the basis of the 138 borings and vibracores which are located on Figure 3 (Eckard et al., 1986; Hayes and Sexton, 1989). The Late Quaternary deposits of the delta plain of the Santee River portion of the delta have been subdivided into a basal fluvial sequence and an upper deltaic sequence. The facies relations of the alluvial and deltaic sequences from the lower delta plain are illustrated in cross-section B-B′ in Figure 4. The extension of the subsurface alluvial valley in the cross-section seaward in a southeasterly direction similar to the trend of the present-day, coarse-grained meander belt is based on regional drill-hole data (Figure 3). The valley sides truncate Late Pleistocene beach-ridge barrier island deposits that occur on the margin of the delta. The alluvial sediments unconformably overlie the Middle and Late Eocene Santee Limestone and Cooper Marl, which have been scoured 10-25 m below MSL. An isopach map of the basal fluvial sand indicates that multistage, stacked, point-bar and channel sands 2-13 m thick have filled part of the incised river channels (see also Figure 5). The laterally continuous floodplain and the freshwater-swamp facies that form the upper 6-10 m of the section contain much finer-grained sediments and generally lack coarse-grained material. This rather abrupt decrease in grain size of the sediments corresponds to a combined decrease in river gradient and increase in tidal influence brought about by the rise in sea level during the Holocene.

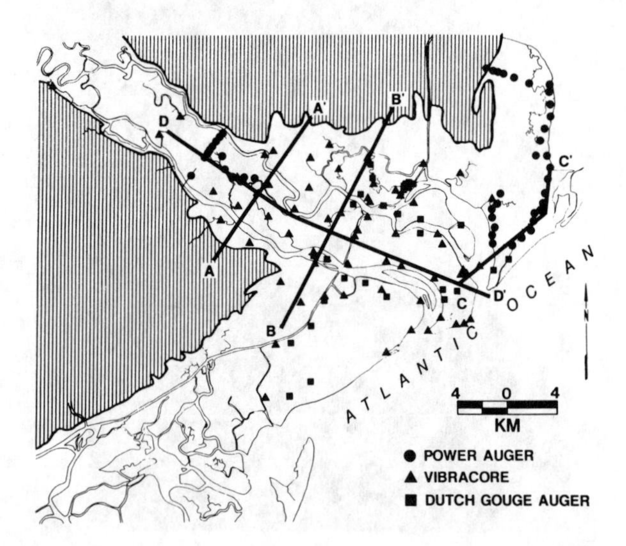

Figure 3. Location of coring stations and stratigraphic cross-sections.

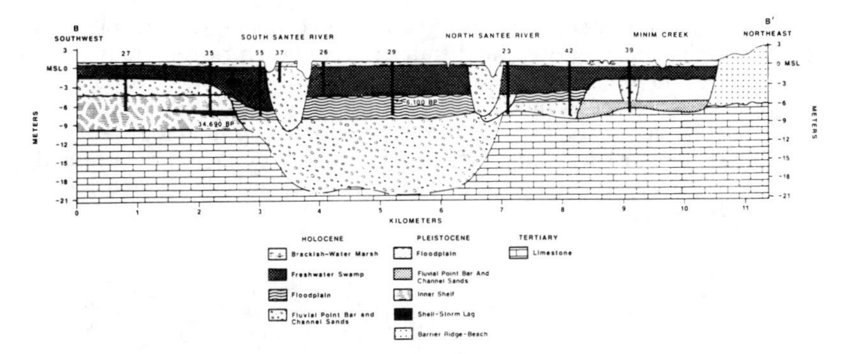

Figure 4. Stratigraphic cross-section B-B′ of the middle portion of the lower delta plain (located on Figure 3).

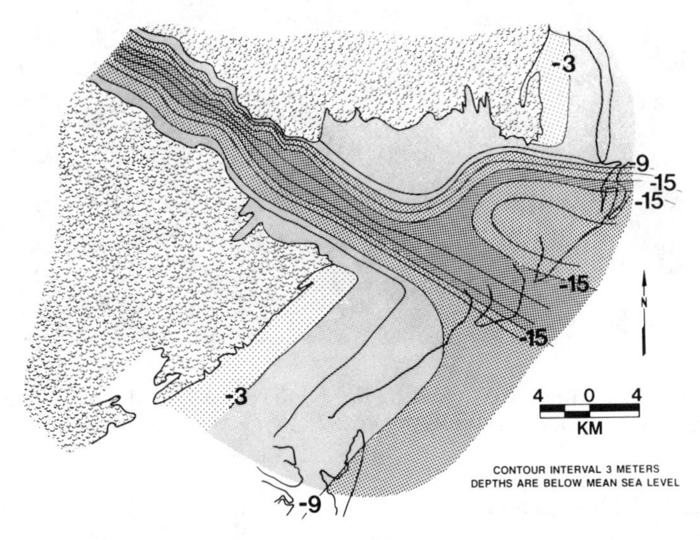

Figure 5. Structure contour map on the Pleistocene lowstand erosional surface which marks the base of the Late Pleistocene-Holocene alluvial valley in the Santee portion of the delta.

Delta-Front Morphology

As can be seen on the satellite image in Figure 2 and the oblique aerial photograph in Figure 6, marine processes have molded the delta front of the Santee/Pee Dee system into a huge, arcuate, sand-dominant bulge in the shoreline. Three major tidal inlets—the Winyah Bay

Entrance, North Santee Inlet, and South Santee Inlet—are separated by recurved spits and regressive and transgressive beach ridges. Large ebb-tidal deltas and small flood-tidal deltas occur at the Santee Inlets. The Winyah Bay entrance is modified by jetties built around the turn of the century. The application of river-dominated delta terminology, such as distributary and distributary mouth bar, to this delta is unwarranted, because the delta-front morphology is almost entirely responsive to marine processes.

Figure 6. Oblique aerial view of the Santee/Pee Dee delta front (color infrared). Photo taken in October 1979 by Peter J. Reinhart.

Delta-Front Stratigraphy

The generalized Holocene sedimentary sequence for the region near the North Santee Inlet (Figure 2) is given in Figure 7. Of course, the delta is positioned over a major alluvial valley, which is illustrated in Figure 5; therefore, the lower portion of the stratigraphic section is alluvial valley fill. The upper part consists of delta-front barrier island sands deposited over freshwater-swamp sediments.

Details of the delta-front stratigraphy are shown in the stratigraphic cross-section given in Figure 8. Fluvial channel and related floodplain sediments were deposited over a lowstand surface of erosion which

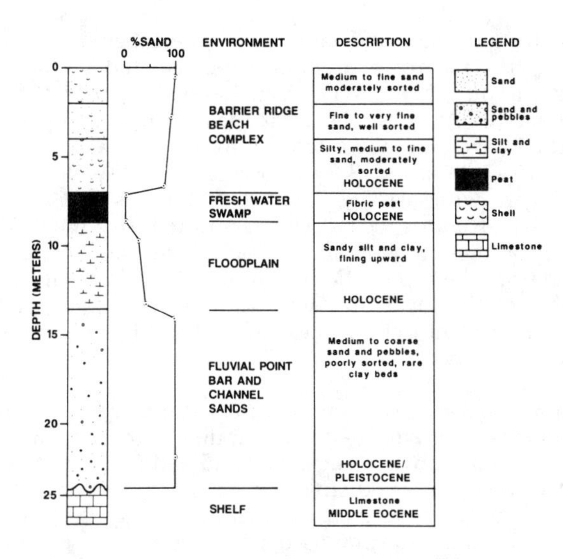

Figure 7. Generalized sedimentary sequence of the delta front near the North Santee Inlet. (From Eckard et al., 1986, Fig. 13).

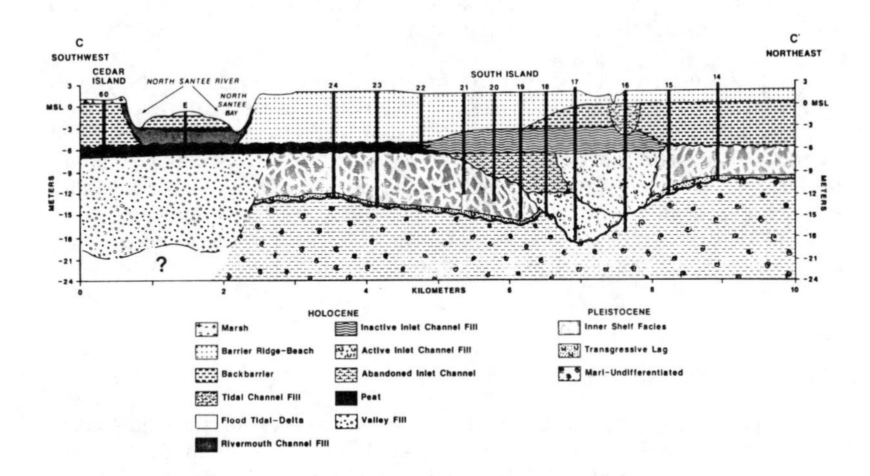

Figure 8. Stratigraphic cross-section C-C′ across the delta front. Located on Figure 3.

truncates a Tertiary marl and late Pleistocene inner-shelf deposits. The marl surface has an erosional topography created in part by alluvial erosional scour during the Wisconsin lowstand [80-17 thousand years before present (B.P.)], with the remaining surface being covered by a coarse sand and shell lag deposited during the Late Pleistocene-Holocene (Flandrian?) (17-7 thousand B.P.) transgression. A well-compacted peat unit interfingers with inactive inlet-fill sediments to the north and is overlain by delta-front deposits. This freshwater peat pinches out 3-4 km landward of the present shoreline beneath lower delta-plain deposits. A peat sample from the top of the unit was carbon dated at 5410 ± 90 B.P. An extension of this peat unit was sampled by Sexton (1987) 1.5 km seaward of the present delta front. Ruby (1981) has reported dates near 4200 B.P. to the south.

A fining-upward, river-mouth/inlet sequence 10 m in thickness, which is composed of coarse- to fine-grained sand and abundant shell fragments, subcrops between auger holes 15 and 21 (Figure 8). The inlet-fill deposit consists of two distinct units: 1) an active inlet channel-fill deposit; and 2) an inactive inlet channel-fill deposit (Tye, 1984). Active channel-fill sediments are composed primarily of coarse- to fine-grained sand mixed with shell material deposited while the inlet remained open, experiencing both riverine and marine depositional processes. The inactive channel-fill sediments include silty fine-grained sand and sandy, silty clay deposited after switching of the inlet position occurred. The muddy, inactive channel-fill sediments were thus deposited in the quiet water of the abandoned inlet (a process discussed by Tye, 1984). Figure 5 illustrates the approximate position of the subsurface channel associated with the river-mouth/inlet deposits underlying Cat and South Islands.

The most recent sediment deposition in the vicinity of the North Santee river-mouth/inlet includes fluvial-estuarine deposits and development of the present flood-tidal delta complex. Hodge (1981) described in detail the sediments and morphology of the flood-tidal delta and underlying sediments at the North Santee River mouth. Mixed fluvial-estuarine sediments comprised of silty, fine- to coarse-grained sand deposited in proximity to the present-day river mouth form the primary deposits, which are overlain by the thin sands of the newly-formed (post-1940) flood-tidal delta.

Offshore of the Delta

Seaward of the subaerial delta lies a complex of transgressively reworked deltaic deposits. The transgressed delta extends 20 km seaward of the active delta front, exhibiting relief in excess of 14 m over a distance of 0.5 km. The sediments of the subtidal transgressed delta deposit, studied by Sexton (1987), range from silt and clay to coarse-grained sand.

Generally, the sediments sampled in the area are mineralogically immature, containing abundant feldspars and rock fragments. Freshwater peat and organic-rich mud have been sampled, confirming that these sediments were part of a delta formed subaerially during lower stands in sea level. Active modern sedimentation also has been observed in the form of landward-oriented, asymmetrical ripples and large-scale, dynamic ridge systems.

Impact of Man

Most river deltas in populated areas have been modified to some extent by man. The Santee/Pee Dee delta is no exception, and in fact, it may be in a class by itself in that regard. Since the beginning of historic times, the upper surface and river flow of the Santee/Pee Dee delta have been modified dramatically by man. From the late 1600s to the late 1800s, extensive rice farming was carried out on the delta, which modified the natural sedimentation and vegetation patterns. The primary impact was the deforestation of the large cypress/tupelo swamp that covered much of the surface of the lower delta plain. The large cypress trees were cut down in the early 1700s, and ponds were formed by dikes for rice cultivation. Presently, many of these ponds are maintained by the U.S. Fish and Wildlife Service, State, and private landowners for wintering waterfowl habitat.

Construction of hydroelectric dams further upriver in the early 1900s steadily reduced sediment supply to the Santee/Pee Dee delta, culminating in 1939, when a large dam was created in central South Carolina that diverted 90 percent of the original discharge of the Santee River into the Cooper River. This diversion oi flow created a major saltwater incursion of the North Santee Channel, and brackish and saltwater marshes grew over sediments deposited in the former cypress/ tupelo swamp. Currently, most of that part of the delta-front shoreline is eroding rapidly, primarily as a result of the decreased supply of sediment to the delta. However, all of the prograded delta-front barrier islands have not, as yet, been eroded away, thus the stratigraphy of many of the barrier islands that remain is regressive in character. There are, however, some transgressive components of the delta front.

In 1986, a variable percentage of the flow of the Santee River above the dam at Lake Marion (not to exceed 40 percent of original discharge) was rediverted back to the river from Lake Moultrie via a newly constructed rediversion canal. This rediversion is not expected to stop the erosional trend at the delta front, however, because the upriver sand supply is cut off by the numerous power dams along the river.

References

Bernard, H.A., 1965. A resume of river delta types (abstract): Am. Assoc. Petrol. Geol. Bull., No. 49, pp. 334-335.

Eckard, T.L., M.O. Hayes, and W.J. Sexton, 1986. Field trip no. 6: deltaic evolution during a major transgression: SEPM Field Guidebook, D.A. Textoris (Ed.), Southeastern U.S. Third Annual Midyear Meeting, Raleigh, N.C., pp. 195-214.

Galloway, W.E., 1975. Process framework for describing the morphology and stratigraphic evolution of deltaic depositional systems: M.L. Broussard (Ed.), Deltas, 2nd Edition, Houston Geol. Soc., Houston, Tex., pp. 87-98.

Hayes, M.O. and W.J. Sexton, 1986. Modern clastic depositional environments, South Carolina; field trip guidebook T371. 28th Int. Geol. Cong., 85 pp.

Hodge, J., 1981. Erosion of the North Santee Delta and development of a flood-tidal delta complex: M.S. Thesis, University of South Carolina, Columbia, S.C., 151 pp.

Price, W.A., 1955. Development of shorelines and coasts. Department of Oceanography, Texas A&M University, Project 63.

Ruby, C.H., 1981. Clastic facies and stratigraphy of a rapidly retreating cuspate foreland, Cape Romain, South Carolina. Ph.D. Dissertation, Department of Geological Sciences, University of South Carolina, Columbia, S.C., 207 pp.

Sexton, W.J., 1987. Morphology and sediment character of mesotidal shoreline depositional environments. Ph.D. Dissertation, Department of Geological Sciences, University of South Carolina, Columbia, S.C., 197 pp.

Tye, R.S., 1984. Geomorphic evolution and stratigraphy of Price and Capers Inlets, South Carolina: Sedimentology, Vol. 31(5), pp. 655-674.

Global Sea Level in the Past and Future Century

Nils-Axel Mörner[1]

Abstract

During the past 150 years sea level may have risen globally by, at the most, 1.1 mm/year from about 1850 to 1930–1940. There is no straight forward relation between climate and sea level. If climate gets warmer, sea level need not at all to rise significantly. If sea level will rise in the near future, it is not likely to rise more than by about 10–20 cm in a century. The deltas of the world are excellent and sensitive recorders of sea level changes and related dynamic processes.

Perspectives

Deltas of the World are excellent natural laboratories for the registration of the complicated interaction between the sea level changes, the supply and accumulation of sediments, the deformation of the sedimentary sequence and the crustal basement, and the meteorological and oceanic coastal dynamics. The investigation and observation of the Deltas of the World, therefore play a very important role for our understanding and estimation of the coastal environmental changes in the near-future.

Past Sea level

In Europe, eustatic sea level rose from about 1830 to 1930–1940 at a mean rate of about 1.1 mm/yr (Mörner, 1973). The origin of this rise is not known. A general rise in temperature is often claimed to be the ultimate cause. From eastern North America, a rise of 2.4 mm/yr is often advocated (e.g. Peltier & Tushinghan, 1989). This may perhaps be correct on a region scale. Such a high rate of sea level rise is impossible

---

[1] Professor, Paleogeophysics & Geodynamics, Stockholm University, S-106 91 Stockholm, Sweden.

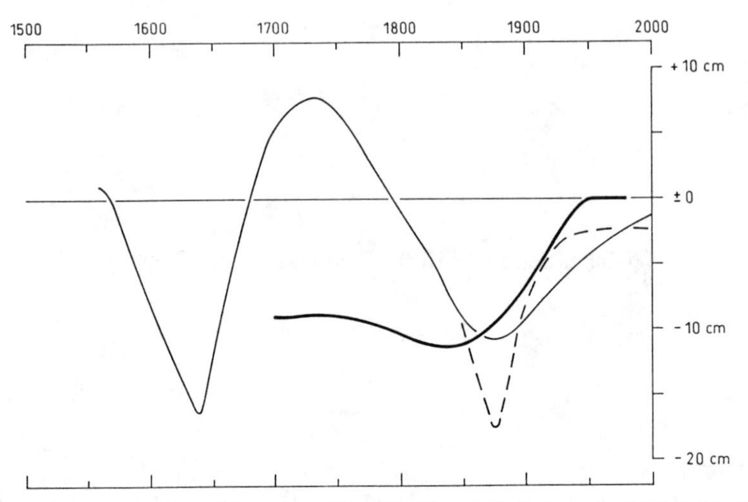

Figure 1. The smoothened mean curve of the "decadal" changes in LOD (with the long-term deceleration trend as zero) recalculated to cm sea level changes assuming that the rotation changes are due to the changes in radius due to sea level changes (thin line). The regional eustatic curve of Northwestern Europe (Mörner, 1973) is added for comparisions (thick line). The two curves are closely similar for the last 150 years (with a 40 year time lag), but totally different prior to 1850. The LOD-changes (thin line) either represent a global eustatic rise of at the most 1.1 mm/year or the effect of mass redistribution in some intermediate oceanic water layer (Mörner, 1992).

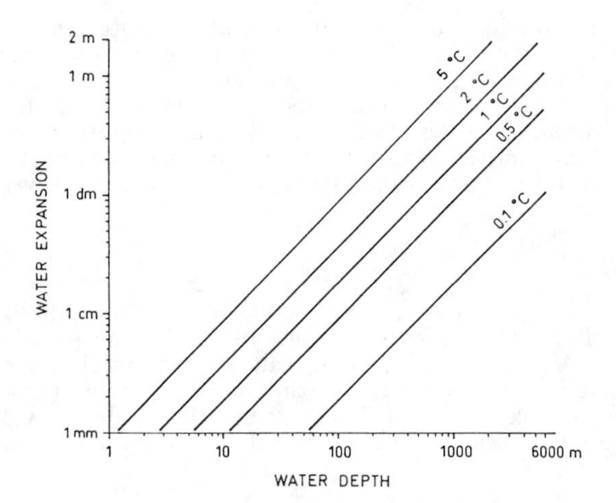

Figure 2. Relations among heating (diagonals) and expansion of the ocean water column. Only the surface water is susceptible to rapid changes in temperature. A present global temperature signal will take about 1000 years to reach the bottom-water. The reaction-times for the intermediate water layes are less well known.

however, on a global scale. This even applies for the analysis by Trumpin & Wahr (1991) arriving at a recent rise of about 1.75 mm/year. In other parts of the world, other figures have been given. The picture is far from uniform, rather the opposite; quite confusing. Furthermore, horizontal redistribution of the water masses seems to dominate over truely vertical (globally equal) changes in the ocean surface level in Late Holocene time (Mörner, 1988).

Our presently available sea level records give a very divergent picture (Pirazzoli, 1986, 1989; Pirazzoli et al., 1989). This is not at all surprising rather what one would expect from the dominance of redistribution of water masses over the globe in Late Holocene time (Mörner, e.g. 1988) and local crustal movements and compaction. Furthermore, the available records primarily represent Northern Hemisphere records.

The best tool for the evaluation of any "global" trend is the analyses of the Earth's rotational record (LOD) because any change in the radius has to affect the rotation, and vice versa. Our analysis (Mörner, 1992) indicates that (1) either there has been a global rise 1830–1930 that amounted to maximum 1.1 mm/yr, or (2) there has been no general rise at all, merely a cyclic redistribution of water masses. This is illustrated in Fig. 1. The importance of this analysis is that it shows that there can have been no global sea level rise larger than 1.1 mm/ /year (which concur perfectly well with the long-term records of the slowly subsiding Dutch coast) in contrast to statements of much higher rates (e.g. Hoffman et al., 1983; Peltier & Tushinghan, 1989; Trumpin & Wahr, 1991; IGCP, 1992).

Climate and sea level

During the maximum rate of degalciation after the 20 Ka glaciation maximum, sea level rose glacial eustatically by about 20–30 mm/year, at the most. These rates represent extreme conditions confined to this period and this process. In Late Holocene time, the rates of eustatic changes must, of course, be considerably smaller. Mörner (1983) estimated that the glacial eustatic component in Late Holocene time does not exceed 1 mm/year.

It is true that climate may get warmer in the near future due to greenhouse effects. But why should this lead to a sea level rise? Sea level may theoretically rise in the near future by means of three mechanisms, all insufficient for a catastrophic "flooding".

(1) *Melting of existing mountain glaciers:* This is not enough because the total water volume in all these glaciers only equals 0.5 m sea level rise (Flint, 1971, Table 4E), and many of them will remain even with a considerable temperature rise.

(2) *Expansion of the ocean water column:* This can only generate some cm to a dm rise (the expansion of the surface water can hardly exceed 10 cm, at the very most 20 cm, and the bottom water can only warm after a time lag in the order of 1000 years, or so). Fig. 2 gives the relations among warming and expansion of the water column.

(3) *Melting of the Antarctic ice cap:* A possible melting is too slow a process to generate any significant effects in the near future (this even applies for a hypothetical West Antarctic surge). The Antarctic ice cap did not melt, but experienced significant expansions both during the

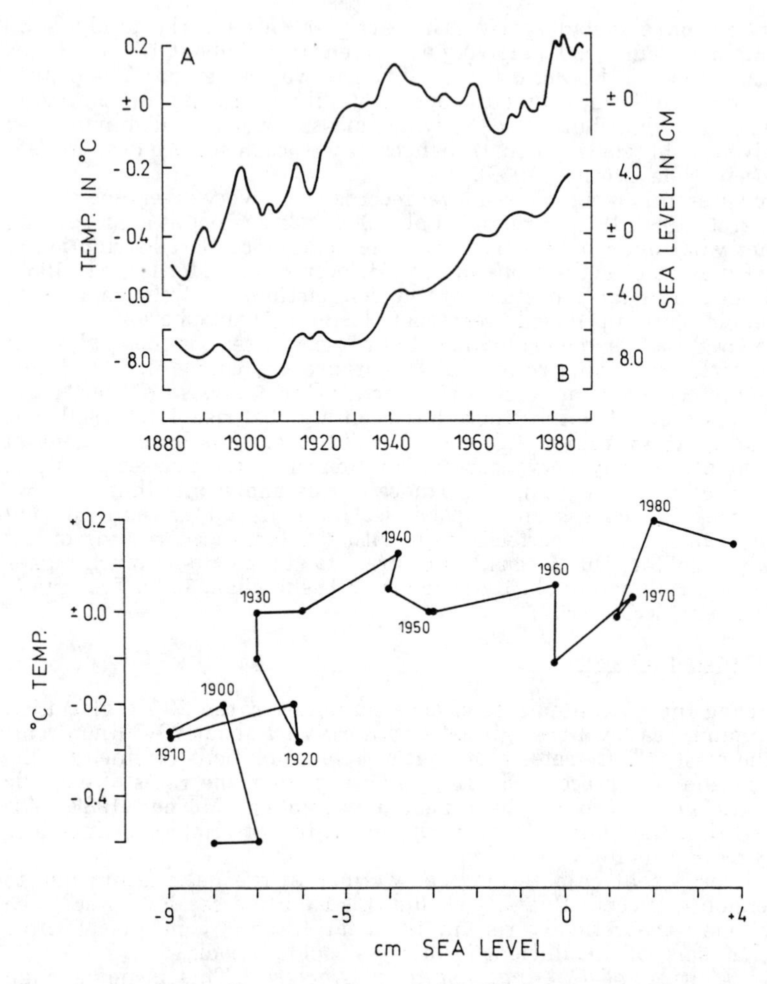

Figure 3. 5-year running mean curves of (A) global mean annual temperature according to Hansen & Lebedeff (1987) and (B) global sea level changes according to Barnett (1988). These curves have been widely used (IGBP 1990) to demonstrate a close correlation between temperature and sea level. This is not at all the case, however, as demonstrated in the lower curve where the temperature values (A) have been plotted against the sea level values (B) for every 5 years. There is actually little or no straight forward correlation between temperature and sea level. When temperature rises significantly between 1890 and 1930, sea level hardly rises at all. When sea level rises significantly between 1930 and 1985, the temperature rise is quite small.

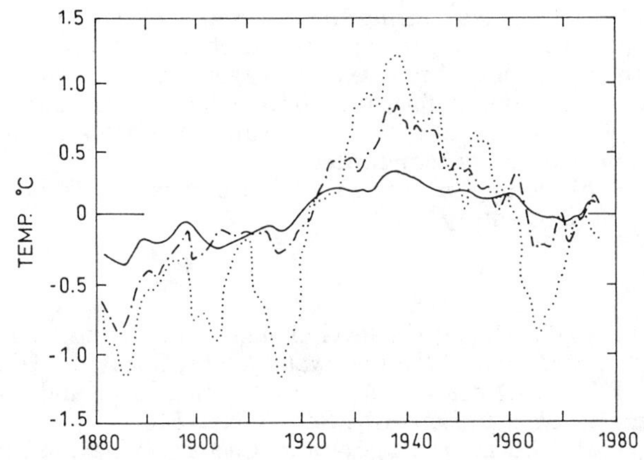

Figure 4. Latitudinal differentiation of observed temperature changes since 1880 (redrawn from WCRP, 1990): 10° N (solid), 65° N (dot-dashed), 80° N (dotted). The differences indicate that latitudinal redistribution of heat plays a fundamental forcing role in climatic changes (as proposed in the model of Mörner, 1988).

Holocene climatic optimum and during the Medieval warm peak.

It has been proposed that there is a close relation between global warming and sea level in the last 100 years (IGBP, 1990, Fig. 7-1). This is in no way well documented. First of all, the curves compared do not substantiate the conclusions drawn. This is illustrated in Fig. 3. Secondly, the sea level curve used does not represent any global trend. The same can be said about the temperature curve used as illustrated by the latitudinal differentiation in Fig. 4.

Future sea level

It is quite clear that the commonly cited values of a hypothetical future sea level rise in the order of as much as 5–30 mm/yr are greatly over-exaggerated (even seemingly more modest values of 3–10 mm/ /year as given by IGBP, 1992). If there will be any sea level rise in the next century, it is likely not to exceed some 10–20 cm. In some low-lying areas, even such a small rise may be quite serious, however. This is, of course, especially true in combination with land subsidence and sediment compaction, effects that often are significant in delta areas.

Conclusions

(1) Deltas of the World are excellent natural laboratories for the registration of the complicated interaction between the sea level changes, the sediment supply and accumulation, the deformation of the sedimentary sequence and the crustal basement, and the meteorological and oceanic coastal dynamics.

(2) The sea level changes during the last centuries do not provide any concordant picture, rather the opposite. Global sea level cannot have risen more than by about 1.1 mm/year from about 1850 to 1930.

(3) These is no straight forward relation between climate and sea level. A global warming will not automatically generate a sea level rise. The relations are for more complicated.

(4) If there will be any future sea level rise, it is not likely to exceed some 10–20 cm in a century.

References

Barnett, T.P., 1988. Global sea level changes. In: *Climate variations over the past century and the greenhouse effect*, Rockville, USA.

Fairbridge, R.W. & Krebs J:r, A.O., 1962. Sea level and southern oscillation. *Geophys, J. Roy. Astr. Soc.*, **6**, 532–545.

Flint, R. F., 1971. *Glacial and Quaternary Geology*. Wiley, 892 pp.

Hansen, J.E. & Lebedeff, S., 1987. Global trends of measured surface air temperature. *J. Geophys. Res.*, **92**, 13345–13372.

Hoffman, J.S., Keyes, D. & Titus, J.G., 1983. Projecting future sea level rise. *US Enviromental Protection Agency*, 266 pp. Washington D.C. Govern. Print. Office.

IGBP, 1990. IGBP Report **12**.

IGBP, 1992. *Global Change: Reducing Uncertainties*. IGBP, 40 pp.

Mörner, N.-A., 1973. Eustatic changes during the last 300 years. *Palaeogeogr. Palaeoclim. Palaeoecol.*, **13**, 1-14.

Mörner, N.-A., 1983. Sea Levels. In: *Mega-Geomorphology* (R. Gardner & H. Scoging, Eds.), p. 73–91, Oxford Univ. Press.

Mörner, N.-A., 1988. Terrestrial variations within given energy, mass and momentum budgets; Paleoclimate, sea level, paleomagnetism, differential rotation and geodynamics. In: *Secular Solar and Geomagnetic Variations in the last 10,000 years* (F.R. Stephenson & A.W. Wolfendale, Eds.), p. 455–478, Kluwer Acad. Press.

Mörner, N.-A., 1992. Sea level changes and Earth's rate of rotation. *J. Coastal Res.*, **8**: 966–971.

Peltier, W.R. & Tushinghan, A.M., 1989. Global sea level rise and the greenhouse effect: Might they be connected? *Science,* 244, 806–810.

Pirazzoli, P.A., 1986. Secular trends of relative sea-level (RSL) changes indicated by tide-gauge records. *J. Coastal Res.*, **SI-1**, p. 1-26.

Pirazzoli, P.A., 1989. Present and near-future global sea-level changes. *Palaeogeogr. Palaeoclim. Palaeoecol.*, **75**, 241–258,

Pirazzoli, P.A., Grant, D.R. & Woodworth, P., 1989. Trends of reletive sea-level changes: Past, present, future. *Quaternary Intern.*, **2**, 63–71.

Trumpin, A.S. & Wahr, J.M., 1991. Constrains on long-period sea level variations. In: *Glacial Isostasy, Sea-Level and Mantle Rheology* (R. Sabadini, K. Lambeck & E. Boschi, Eds.), NATO C-334, p. 271–284. Kluwer Acad. Publ. Press.

WCRP, 1990. World Climate Research Program. *Global Climate Change*. WMO & ICSU, 35 pp.

Louisiana Cheniers: Clues to Mississippi Delta History

William F. Tanner[1]

Abstract

The cheniers of southern Louisiana have map differences, bedding differences, and granulometric differences that distinguish them from swash-built beach ridges and from dune ridges. Instead, they are like late-storm settling deposits in near-shore water beyond the surf zone.

Granulometric methods (suite statistics) distinguish between these cheniers and swash-built ridges. On a plot of mean kurtosis vs standard deviation of kurtosis, these cheniers are far from all known beach ridges or beaches. Mean K is close to 10 (very high) in a region where 4 was expected. The standard deviation of the kurtosis is also about 10, where 2 or less was expected. No other beach ridge set is known with this parameter as high as six.

Observed bedding was parallel and convex-up, like that in large storm-surge ridges in other places. But the Louisiana cheniers differ from large storm-surge ridges, many of which are 6-9 m high and represent long intervals of upward growth; the Louisiana examples are spread out.

In map view, many cheniers have splayed ends, showing that they are composites of narrower ridges.

The cheniers under study were built by settling over periods of a few centuries when sea level stood 1-2 m higher than now, but not by a single storm. Each wide swale between cheniers represents a similar interval when sea level stood 1-3 m lower than now.

Each chenier may be equal to a beach ridge <u>set</u> in other settings, and thus may reflect some of the parts of late Holocene sea level history. The type of sediment indicates proximity to a major river delta.

Whether or not a soft sub-strate is present, these Louisiana examples can be diagnosed strictly on the basis of granulometry and inferred hydrodynamic processes.

----

[1] Regents Professor, Geology Dep't B-160, Florida State Univ., Tallahassee, Florida 32306-3026 U.S.A.

Introduction

Beach ridges of the swash-built type are typically found in systems each of which may contain up to 200 or more individual ridges. The oldest presently-known more-or-less continuous sequence of beach ridges of this type dates back to about 12,000 B.P.; these ridges are located in a region of slow glacio-isostatic rebound (northernmost Denmark; Tanner in press). In more-or-less stable areas, such as along the coasts of the Gulf of Mexico, swash-built beach ridges date back only to about 3,000-3,200 B.P. Therefore the Danish ridges are the standard for middle and late Holocene time.

Swash-type beach ridges are typically built at intervals of a few decades. Two ridge systems are known where the interval was three or four years, three more where it was about 11 years, but in general it was 30 to 60 years. For example, a system of 60 ridges with a mean spacing of 50 years provides a reasonably detailed history of that coastal strip, covering 3,000 years.

Each narrow ridge-and-swale in such systems represents a small sea level rise-and-drop, commonly in the range of 5-50 cm. This is generally the only detectable and persistent periodicity in the system. The cheniers are much wider than those narrow ridges.

Many beach ridge systems were built well away from any important delta; examples are known from Florida (Cape Canaveral; Cayo Costa Island; Sanibel Island; Marco Island); Mexico (Isla del Carmen; Teacapán); Venezuela (Golfo de Venezuela); Canada (Richmond Bay, Quebec); Denmark (Skagen; Jerup; Tversted); Germany (Geltinger Birck; Schleimünde; Heiligenhafen); etc. Other systems were built immediately adjacent to or down-drift from certain major river deltas, such as the Niger, Río Paraíba, Usumacinta (Mexico), Río Grande, Apalachicola (U.S.), Alabama (U.S.) and Mississippi.

Well-developed systems are composed of beach ridge sets, each set having about five, to perhaps 25 or more, narrow ridges in it. Adjacent sets commonly stand at different elevations, with a height difference of a meter or two. High sets were formed at times of high sea level position, low sets when sea level was lower. Sand grain size parameters, using suite statistics methods (Tanner 1991-a), also indicate either high or low sea levels, and this relationship can be confirmed in places where set height differences are obvious.

Geometry

In Louisiana, a chenier is commonly a kilometer or so wide and splays near the ends and on the landward side into many narrow ridges which are like the swash-built or settling-lag ridges on other well-studied beach ridge plains. Common map spacings for ridges in other places are 30-60 m., locally less. Twenty ridges with a mean spacing of 50 m makes a set one kilometer wide. Individual ridges and swales are hard to identify away from the splays, but one may note that the combination is a composite of ridges. This composite should be matched with ridge <u>sets</u>, rather than with individual ridges in other areas. Because each narrow ridge (in other areas) commonly required 30-60 years to form, such a set may represent centuries, rather than decades or years.

Relief on Grand Chenier, in the vicinity of the town of that name, in Louisiana, is about 2-3 meters, which is much too large for ordinary swash-built beach ridges. Instead, this is an acceptable value for the difference in elevation between ordinary ridge sets.

Von Drehle (1973) studied composites in Gulf County, Fla. (near the town of Port St. Joe), and obtained measurements much like those on the cheniers: spacing (crest to crest) of 400-1,200 meters, wide low swales between the ridges (or composites), and little or no visible relief on the ridge tops. However, his ridges were built far from a delta, and therefore have granulometric parameters which differ greatly from those of the cheniers: he found, for example, mean kurtosis to be only 3.626 (for 101 samples) and the standard deviation of kurtosis to be about 0.23. These are numbers that one might obtain on a low-to-moderate energy beach made of clean sand.

In the Louisiana chenier district the same logic can be applied to what appear to be very wide swales: they may be low sets of individual ridges the details of which cannot be seen very well, perhaps because low positions have facilitated settling of additional sediment at a later time, thus masking the original map geometry.

Therefore cheniers should be studied in comparison with ridge <u>sets</u> which are well defined in certain other places such as on St. Vincent Island, near the town of Apalachicola, Florida, where sets typically include 5 to 20 or more ridges and which required two to eight or nine centuries for construction.

<u>Field Work</u>

Field work in the chenier country of southern Louisiana has been carried out at irregular intervals over a period of years. This has included inspection of pits in the cheniers, in which parallel convex-upward bedding, rather than beach-type cross-bedding, could be seen.

Sampling for this one report was done by Barry and Lynn Reik, in the Spring of 1991, near the town of Grand Chenier, located on Louisiana State Highway 82, in Cameron Parish (= County). Twenty-four samples were collected, on a north-south line (fig. 1), offset so that marshy areas could be avoided. The ridges that were traversed on the sample line represent a high sea level 1,300-1,800 B.P. (C-14 dates: Gould & McFarlan 1959).

The profile crossed two wide ridges, plus one wide swale; the latter stands relatively low. Any very wide swale may be a low set of ridges, partly masked by later sediments hence should be sampled for continuity reasons. Although a dry season was chosen for field work, a high ground water position made sampling difficult.

Each sample was approximately 100 grams, taken from a depth of about 30 cm. Thirteen of the samples were selected for detailed processing.

Granulometry: Suite Statistics

Samples were treated with HCl, washed, dried, split once, and sieved for 30 minutes in a nest of quarter-phi screens (Socci and Tanner 1980). Fractions were weighed to 0.0001 grams. Raw weights were processed in the computer program GRAN-7, which produces the first six moment measures, numerous other parameters, a histogram and a probability plot. These moment measures are the mean (as a substitute for the first moment, which is zero), the standard deviation, the skewness, the kurtosis, the fifth moment measure and the sixth moment measure. None is a graphic measure. Each one represents the sample fully (Tanner 1991-a, 1991-b).

Most sandy beach ridges have granulometry like the nearby fair-weather beach, but not like the storm beach. The Louisiana cheniers are distinctively different at this point, and do not match either kind of beach.

Suite mean parameters are as follows (phi scale): mean 2.508 (0.176 mm), standard deviation 0.627, skewness -0.007, kurtosis 10.339, Fifth -25.405 and Sixth 403.883. The standard deviations of the first four parameters are 0.538, 0.158, 1.533 and 9.648 (kurtosis). The suite mean and standard deviation of the kurtosis are noteworthy (10.339, 9.648): this is a most unusual pair of values.

The suite mean skewness indicates a beach environment of deposition. The standard deviations of the means and of the standard deviations indicate beach or settling from water. The relative dispersions of the skewness and kurtosis indicate settling from water, as does the geometry of the probability plot. The sixth moment measure, plotted against kurtosis, indicates settling from water. The mean and standard deviation of weight percent of the tail of fines (4 phi and finer) suggest a beach, but this is not a good example. Relative dispersions of the mean and of the standard deviation likewise suggest a beach, but the example is not very good. Overall the indication is a beach (perhaps not well developed) plus a major contribution by settling from water (Tanner 1991-a, 1991-b).

Granulometry: Kurtosis

The grain-size kurtosis when inverted shows clearly the wave energy content, degree of settling from water,

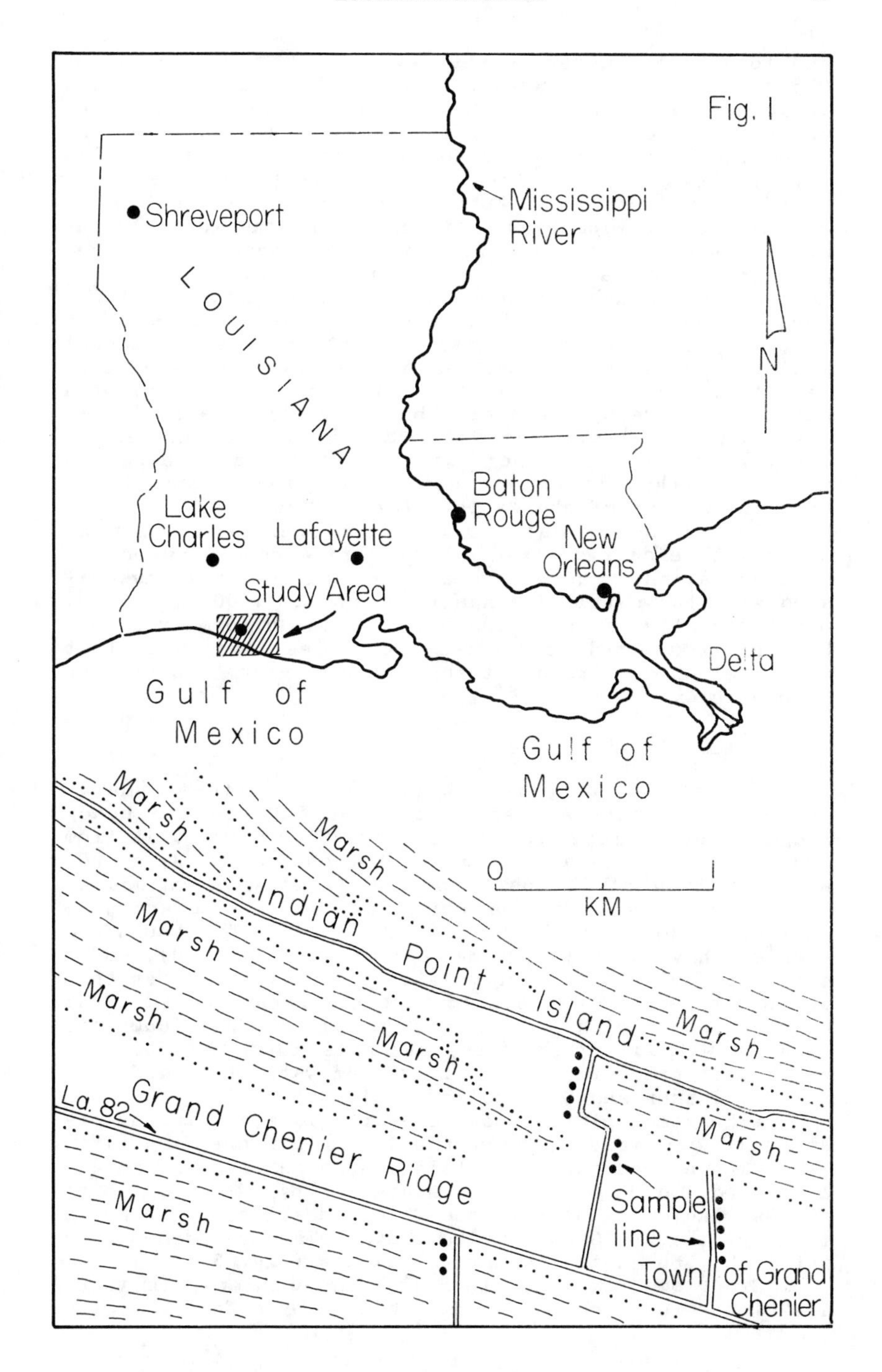
Fig. 1
Shreveport
LOUISIANA
Mississippi
River
N
Lake
Charles
Lafayette
Baton
Rouge
New
Orleans
Study Area
Delta
Gulf of
Mexico
Gulf of
Mexico
Marsh
Marsh
0
1
KM
Indian Point Island
Marsh
Marsh
Marsh
Marsh
Marsh
La. 82
Grand Chenier Ridge
Sample
line
Marsh
Town of Grand
Chenier

and long-term changes in sea level in the range of about 0.5 to 5 m (Tanner 1992-a). Individual kurtosis values indicate near-shore wave energy density and also show the extent to which settling-from-water was important in construction of the ridge. Kurtosis close to 3 (the Gaussian value) shows moderate-to-high wave energy density, with little or no settling. Kurtosis close to 4 indicates low energy without important settling. As the kurtosis increases above 4, the settling component becomes more and more important; at values of 10 and higher, settling was dominant and swash action on the beach was minimal. Long-term changes in kurtosis, covering four or more beach ridges (therefore, in general, a century or more) cannot indicate tides or storms, but rather identify long-term sea level changes (Tanner 1992-b). This is because, on a faintly concave-up transverse bottom profile, a small sea level change alters the water depth in the near-shore zone, hence modifies the degree to which wave energy is drained to the bottom before the wave breaks, and therefore the amount of energy left for the surf.

Granulometric data from the younger parts of many different beach ridge plains (from five countries on two continents) show the sea level drop and then rise associated with the Little Ice Age (from about 1,200-1,300 A.D. until recently; Tanner 1992-b). This uniformity of behavior, from widely-scattered localities, indicates that sea level change, rather than some short-term factor, was responsible.

Granulometry: Cross-Plots

A plot of suite-mean kurtosis vs suite-standard-deviation of kurtosis shows that most beach and beach ridge sample suites extend in a fairly narrow band from low values on both axes (mean of kurtosis about 3, standard deviation of kurtosis about 0.3) to high values on both axes (mean of kurtosis perhaps 10-20, standard deviation of kurtosis about 10; fig. 2). The Louisiana cheniers have a very high mean kurtosis (about 10), and a standard deviation of kurtosis that is higher by far than any other beach ridge system that has been studied (9.5; no other beach ridge set as high as 6). These numbers, with or without other data, indicate that the cheniers were built primarily by settling-from-water (the model of Postma; Tanner and Demirpolat 1988).

Likewise a plot of mean-of-kurtosis against mean-of-standard deviation places the Louisiana cheniers with other known products of settling, such as the horizontally-bedded settling-lag (not swash-built) ridges on Mesa del Gavilán, west of the Boca Chica locality on the north flank of the Río Grande delta (Texas; Tanner and Demirpolat 1988), but not with mature beaches (fig. 3).

Granulometric parameters for the cheniers, in fact, are closer to those for sands produced by offshore sett-

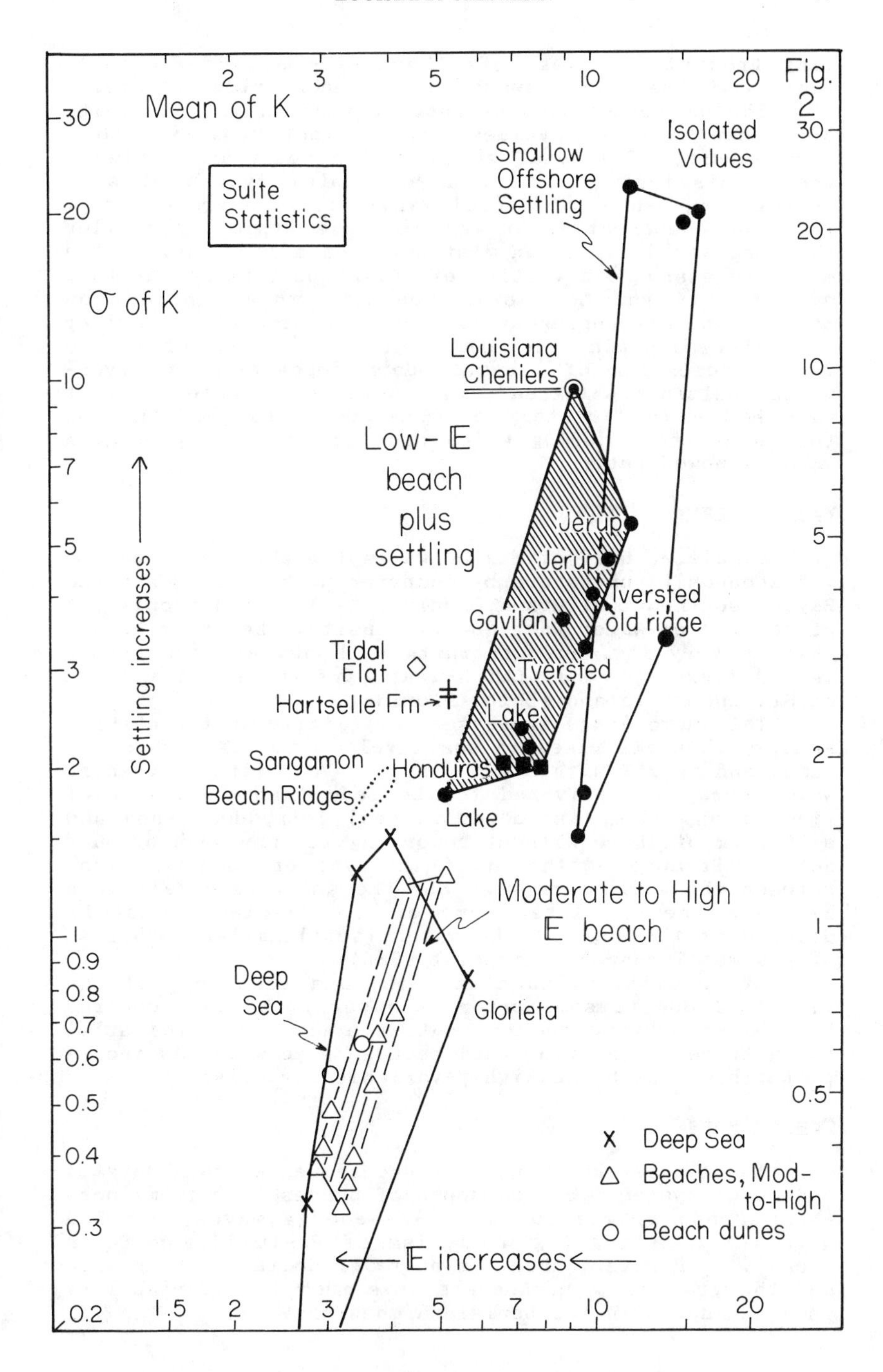
Fig. 2
Mean of K
Suite Statistics
σ of K
Settling increases
Shallow Offshore Settling
Isolated Values
Louisiana Cheniers
Low-E beach plus settling
Jerup
Jerup
Tversted old ridge
Gavilán
Tidal Flat
Tversted
Hartselle Fm
Lake
Sangamon Beach Ridges
Honduras
Lake
Moderate to High E beach
Deep Sea
Glorieta
Deep Sea
Beaches, Mod-to-High
Beach dunes
E increases

ling from storm waves (off the east coast of Florida), than to those from swash-built beach ridge systems.

The extremely high standard deviation (of the suite of chenier kurtosis values) may reflect proximity to a large source of fluvial sediment (in Louisiana, this is the Mississippi River, with a complex system of sub-deltas, and hence much local variability in grain size). The fine component (silt and clay) is typically smaller than one would expect in most parts of a large delta: the mean and standard deviation of weight-percent in the tail of fines (4 phi & finer) are like those indices on beaches and may represent late modification of an earlier fluvial-type grain-size distribution. However, there is no indication of direct modern deposition of river-mouth sediments. Instead, the former river material must have had a long history of back-and-forth shuffling on the inner shelf during which most of the fine particles were winnowed out.

Wave Energy

Simulated Gulf of Mexico waves (height = 2 m, period = 5 seconds), run ashore by computer methods at Constance Bayou (southeast of Grand Chenier, La.), retain only 6 % of their deep-water wave energy density, when they break. This is very low, and contrasts with 50% at South Padre Island (Texas), and with 38% at Captiva Island and 23% at St.Vincent Island (both, Florida).

This percentage increases very little in the chenier area, with small rises in sea level: to 14% with a 1-m rise, and to 23% with a 2-m rise. Very little offshore wave energy is delivered to the beach, and a sea level rise of any kind (including surges) introduces sand and silt from offshore without reworking it very much by surf action. Because of the brief interval of no-net-motion, between the short-term rise and the short-term fall in a single surge, at least some of this sediment must be dropped at the edge of the water (settling-lag mechanism of Postma; Tanner & Demirpolat 1988).

At deposition, such sand (a) has lost most of the fine tail due to some prior wave work, (b) has acquired beach-like sorting due to that wave work, but (c) still has a trace of the fine tail because wave-work was incomplete (hence high and highly-variable kurtosis).

Comparisons

The longest known beach ridge sequence, the Tversted and Jerup systems located south of Skagen in extreme northern Denmark, was built by low-energy waves, coupled with the settling-lag mechanism of Postma (Tanner, in press). However, there is no river delta in the region and the grain size parameters have much less variability than the sands in the Louisiana cheniers.

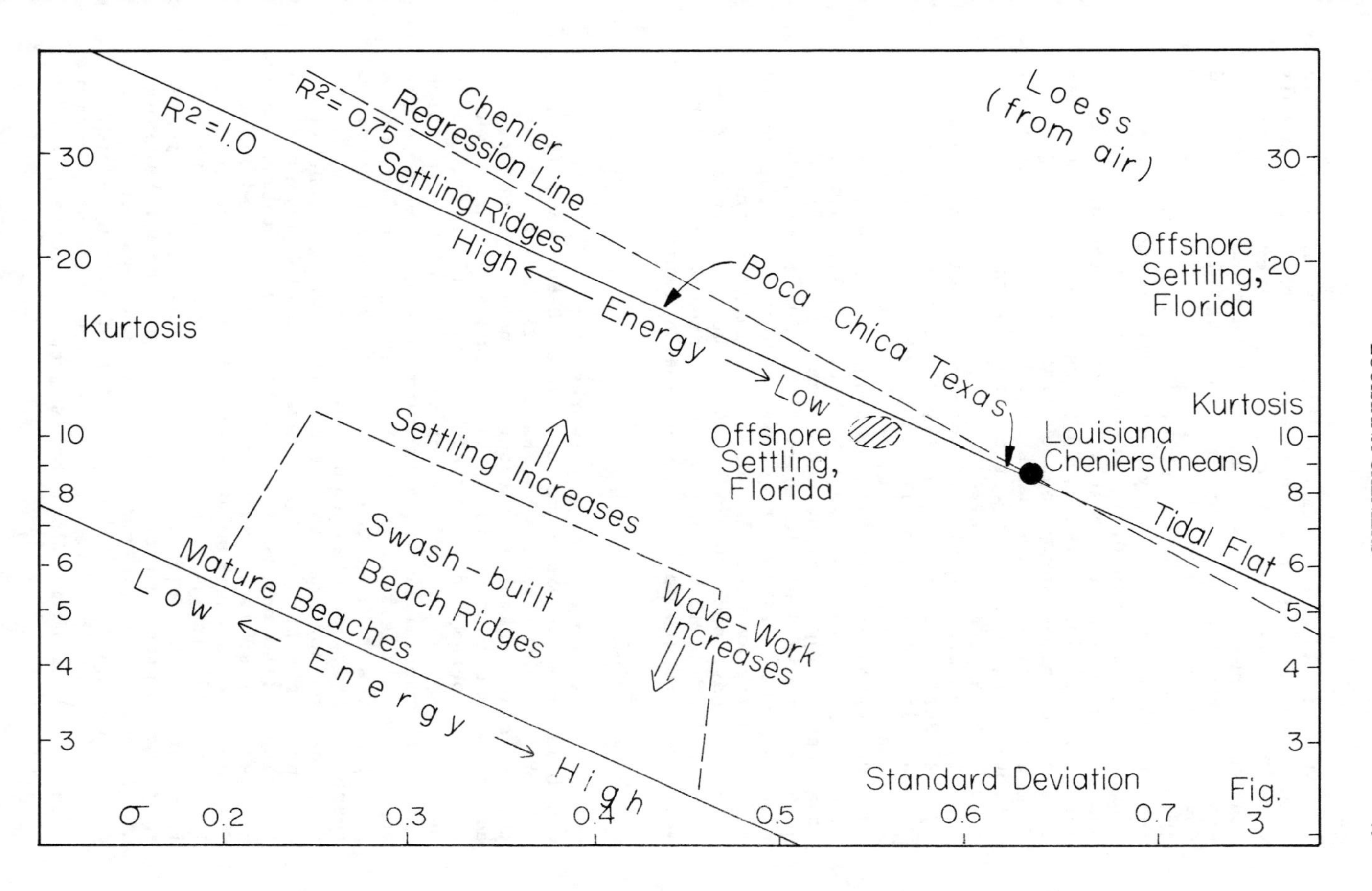
Chenier Regression Line
R2 = 0.75
R2 = 1.0
Settling Ridges
High ← Energy → Low
Loess (from air)
Offshore Settling, Florida
Boca Chica Texas
Kurtosis
Kurtosis
Offshore Settling, Florida
Louisiana Cheniers (means)
Tidal Flat
Settling Increases
Swash-built Beach Ridges
Mature Beaches
Wave-Work Increases
Low ← Energy → High
Standard Deviation
σ
0.2
0.3
0.4
0.5
0.6
0.7
30
20
10
8
6
5
4
3

Fig. 3

The St. Vincent Island (Fla.) beach ridge system is located on the rim of the Apalachicola delta, and is only a few kilometers from the mouth of the river. These sands (taken as one suite) have moderate wave-energy kurtosis and only modest evidence for settling: they were laid down primarily by swash. But the grain size parameters change in a systematic fashion from the oldest ridges (about 3,000 B.P.) to the youngest. The former have granulometric indicators of river transport, and the latter, swash work on a mature beach. Therefore the delta sands were reworked during the 3,000 years to have characteristics of moderate-energy beaches (Tanner 1992-b).

How is it that the sands of the Grand Chenier area, in Louisiana, do not show the same reworking? Several possibilities can be listed:

1. They are about 1,500 years old: not enough time.
2. Surf-zone energy levels, there, are very low.
3. Older delta deposits, exposed on the inner shelf, are much more varied than in the Florida example.

Construction rates for the cheniers have been close to 2-5 m/year, in comparison with 0.9-to-1.2 m/yr in most other beach ridge plains, hence very fast. A good count of narrow individual ridges is difficult to make, but the time interval between them may have been close to 10 years, on the average. In most beach ridge plains, common intervals have been 30 to 50 years, but intervals of 3-4 years and of 10-12 years have been noted at a few places where seaward growth has been unusually rapid.

Sea Level Change

A single ridge does not represent a single storm but was built over an interval of about 30-60 years (Tanner 1992-a, 1992-b, in press). Each single, narrow ridge in the Louisiana examples is thought to have been constructed as a result of settling from occasional high water positions, perhaps at intervals of about 10 years. Sandy cheniers may be composites, or sets of ridges, in which case each chenier covers a century or more, the adjacent swale covers a similar time span, and each represents a position of sea level that lasted longer than a storm, a season, a year, or even a decade.

The concept that each ridge was built in one storm does not apply to beach ridge systems, which may have 50-150 ridges (the narrow swash-built type) in them. Six storms per year is a reasonable and widely cited figure. This count requires that 60 (narrow) ridges were built in 10 years, 120 (narrow) ridges in 20 years, and even the long Jerup system (154 ridges, 7,800 years), in 26 years. Therefore almost no beach ridge plain could have any significant cover of mature trees, and essentially all beach ridge growth would be a matter of modern record.

This poor logic would also require that each chenier (composite) was built in 30 or so years, contrary to the available dates.

The long record (about 12,000 years) in the Tversted and Jerup systems (in northern Denmark), with data points every 30-50 years, shows detailed history of various sea level changes in the range of 0.5 to about 5 m. In this historical sequence, the initiation of Mississippi River sub-deltas and of Louisiana cheniers is associated with sea level rises.

Until a more detailed history of cheniers has been produced it is thought that the Tversted-Jerup record can be borrowed, to provide a temporary substitute for use in the Mississippi River delta region. Perhaps the Jerup-Tversted record will be superior to any improved history of the cheniers of Louisiana, because of the extreme variability of the settling component in the latter. In any event, the Tversted-Jerup data provide a framework within which it is profitable to study Louisiana cheniers and Mississippi River delta history in more detail.

The Narrow Ridge

In many other beach ridge plains, like the Jerup and Tversted systems, where individual ridges are reasonably sharply defined, the map spacing is commonly 30 to 60 m, the time interval is 30 to 60 years, and the accretion rate was roughly 1.0 m/yr. This kind of detail is very difficult to see in the cheniers. If one examines the splayed ends and some of the back sides, the results may not be satisfactory, because experience elsewhere shows that splaying and curvature do not preserve the features of the individual ridges very well.

Therefore we look to other areas for details. Each narrow swash-built ridge appears to have been built by a pair (couplet) of small sea-level changes in the range of perhaps 5 to 40 cm. This is one rise, and one fall.

The transverse bottom profile from the beach seaward is typically gently concave up. This means that a small rise in sea level changes the water depth at any selected distance from the ambient water's edge. This in turn changes the energy drain as the wave approaches the surf. As a result, a small rise in sea level causes an increase in breaker-zone energy, a higher run-up of the swash, and deposition to a higher point on the beach. If this higher position persists for a few years, a beach ridge will be built.

The next small drop in sea level causes a decrease in surf zone energy and less run-up. If this new position persists for a few years, a narrow swale will be built. Detail of this kind represents very fine structure, which provides considerable insight into short-term (decadal) behavior of sea level, and hence presumably of the world climate. This sequence of events is easy to demonstrate with computer simulation methods, using known bathymetry and appropriate sea level changes.

However, for Mississippi River delta history, it is too much detail. But the chenier is a composite, and like a beach ridge set, it depicts sea level changes at the century scale: two, three, four or more.

The resulting pattern is small changes (a few years, or decades) on top of larger changes (centuries; one to four or five meters), and the latter may correlate -- at least in a general way -- with delta events, such as the initiation of new sub-deltas.

From the stand-point of general geological history, it can be noted that beach ridge sets do not have any obvious periodicity, so that the scheme that was used above (decades, centuries) cannot be extrapolated in rigorous numerical fashion to larger and larger values, although longer, to much-longer, intervals are known. That is, it appears that almost any interval, longer than a century, may be possible. Therefore one does not expect an orderly calendar arrangement of cheniers such that we can extrapolate into time spans where little or nothing is known.

Delta History

As a composite, built over an interval of centuries, each major chenier appears to represent a time of late Holocene high sea level position, and each wide swale, a time of low position. The initiation of Mississippi River sub-deltas coincides, in a general way, with sea level rises between low stands and ensuing high stands (Tanner 1991-c). This association may be due to the following:

1. During sea level rise, deltaic deposition should be shifted landward, burying previous delta deposits and creating a sediment body which will not be reworked, or destroyed, by waves during a subsequent sea level drop. Therefore there may be a high potential for both preservation and exposure (for later inspection and sampling).

2. After sea level rise is terminated, the new sub-delta should prograde, so that younger parts of the sediment body represent the high stand rather than the rise.

3. The match may not be perfect, because an equilibrium of some kind must be considered: rate of sea level rise, vs volume delivery of river sediment. If sea level rise is too fast or too slow for the delivery rate, the association stated here may not hold.

Substrate

Cheniers have been defined at various times as beach ridges which rest on a soft substrate, such as clay or peat (cf. Gary et al 1973; Otvos and Price 1982).

In the present paper, the position is taken that Louisiana cheniers were built basically by a different process, specifically by settling from high water rather than by energetic swash on a sandy beach such as tourists like to visit. Whether or not there is a soft substrate is not germane to this discussion, which is fundamentally hydrodynamic in nature. The presence of the soft peat or clay which has been shown to underlie some cheniers is consonant with the near-delta location: settling of very fine particles to form a clay layer, prior to deposition of the chenier. Such a soft layer is not necessary.

Conclusions

The individual chenier appears to be a composite of many narrow (typical) ridges and swales: e.g., a ridge set. Such narrow ridges commonly have map spacings of 30-60 m, accretion rates of about 1 m/yr and time intervals of 30-60 years. In the Louisiana chenier district, map spacing of the narrow ridges is about the same, but accretion rates are much faster than normal, and time intervals (approximately 10 years) are shorter. Based on experience in many other areas, sets (composites) should be hundreds of meters wide, or more than a kilometer wide, and should represent a century, or more, of time. With this interpretation, chenier history can be matched against history of other beach ridge plains, such as the Tversted and Jerup low-energy swash-built systems in extreme north Denmark, where the record is 11,500-12,000 years long, and is clear, detailed and stable.

Results obtained from "suite statistics" analysis of chenier sands are important in the environmental interpretation of these deposits. The suite-mean value of the grain-size kurtosis, and the suite-standard-deviation of that kurtosis (key parameters for environmental interpretation) indicate settling from water.

Sediments in the cheniers that were sampled indicate a history of low-energy wave work plus important settling from water. The surf zone energy was low, for this part of the Gulf, because only a very small fraction of open-sea wave energy was delivered to the coast. Settling effects were dominant; chenier sands are much like modern late-storm sediments collected on the inner continental shelf, outside the surf zone, off the coast of Florida, and much like the settling deposits near Boca Chica, Tex.

Like ordinary narrow swash-built beach ridges, each individual ridge in the composite cheniers was not made by one individual storm, but rather over an interval of decades. The chenier, therefore, represents centuries.

Chenier sediments, in the study ridges, are thought to be reworked Mississippi River deposits, from two or more old sub-deltas, modified somewhat by later back-and-forth shuffling on the inner shelf, and finally laid down on the coast, by settling from high water.

A chenier is not simply a swash-built beach ridge, resting on a soft sub-strate (clay or peat). The nature of the substrate is of no importance to the argument: the chenier owes its distinctive characteristics to distinctive hydrodynamic processes.

References

Gary, M., R. McAfee jr., and C.L.Wolf, 1973. Glossary of Geology. American Geol. Inst., Washington D.C.; 805 p.

Gould, H.R., and E.McFarlan, 1959. Geological history of the chenier plain, southwestern Louisiana. Trans. Gulf Coast Assoc. of Geological Societies v. 9 p. 261-270.

Otvos, E.G., and W. Armstrong Price, 1982. Chenier and chenier plain. Pp. 206-207, in: Encyclopedia of Beaches and Coastal Environments, M.L. Schwartz, ed.; Hutchinson Ross; Stroudsburg, Penn. U.S.A.; 940 p.

Socci, A., and W.F.Tanner, 1980. Little-known but important papers on grain-size analysis. Sedimentology, vol. 27 p. 231-232.

Tanner, W.F., 1987. Spatial and temporal factors controlling over-topping of coastal ridges.In: Flood Hydrology, V.P.Singh ed.; D.Reidel Publ. Co., Dordrecht; 429 p.

Tanner, W.F., 1991. Suite statistics: The hydrodynamic evolution of the sediment pool. Pp. 225-236 in: Principles, methods and application of particle size analysis, J.P.M. Syvitski ed.; Cambridge U.Press, Cambridge; 368 p.

Tanner, W.F., 1991. Application of suite statistics to stratigraphy and sea-level changes. Pp. 283-292, in: Principles, methods and application of particle size analysis, J. P.M.Syvitski, ed.; Cambridge U. Press; 368 p.

Tanner, W.F., 1991.The "Gulf of Mexico" late Holocene sea level curve and river delta history. Trans., Gulf Coast Association of Geological Societies, vol. 41, p. 583-589.

Tanner, W.F., 1992. 3000 years of sea level change. Bull., Amer. Meteorological Soc., v. 73 p. 297-303.

Tanner, W.F., 1992. Late Holocene sea-level changes from grain-size data: evidence from the Gulf of Mexico. The Holocene, vol. 2, p. 249-254.

Tanner, W. F., in press (1993). An 8,000-year record of sea level change from grain-size parameters: data from beach ridges in Denmark. The Holocene, vol. 3.

Tanner, W. F., and S. Demirpolat, 1988. New beach ridge type: severely limited fetch, very shallow water. Trans. Gulf Coast Assoc. of Geol. Societies, v. 39 p. 553-562.

von Drehle, W.F., 1973. A sedimentary investigation of the large, linear sand bodies in the Gulf County, Fla., Canal. Unpubl. thesis, Fla. State U., Tallahassee; 119 p.

Humid tropical zone deltas, their potential and limitations for development: examples from Java, Indonesia

Piet Hoekstra[1]

Abstract

Monsoonal rivers on Java are responsible for the rapid growth and development of river deltas along the north coast of the island. The climatic and local environmental conditions in these deltas result in a range of problems with concern to water quality and quantity, the excessive supply and deposition of sediment and the present land use. Measures already taken to improve the development opportunities of the deltas include methods to control the sediment load of the rivers. By introducing river training works inside the deltas and by increasing the degree of planning in an early stage, a further improvement of the local conditions is expected.

Introduction

River deltas in the humid tropical zone are frequently facing specific problems which, for a substantial part, are related to characteristic climatic and environmental conditions. High intensity rainfalls in combination with intense tropical weathering of rocks and soils commonly result in high denudation rates and severe erosion in river basins and catchment areas. Large volumes of water and sediment are carried towards the seas. Few of these tropical rivers and their deltas have been studied in appreciable detail, even though it is known that tropical rivers, especially in the humid tropical zone, are responsible for about 75% of the total annual discharge of suspended matter on earth (see Fig. 1; Jansen, 1979; Coleman, 1982 and Milliman and Meade, 1983). In many cases the large supply of sediment results in a rapid progradation of the river deltas, creating new opportunities for development. The reverse side of the medal is though that the same river regime and sediment supply also

---

[1]Netherlands Centre for Coastal Research (NCK), Institute for Marine and Atmospheric Research, Utrecht University, P.O. Box 80.115, 3508 TC UTRECHT, the Netherlands, Fax +31-30-540604

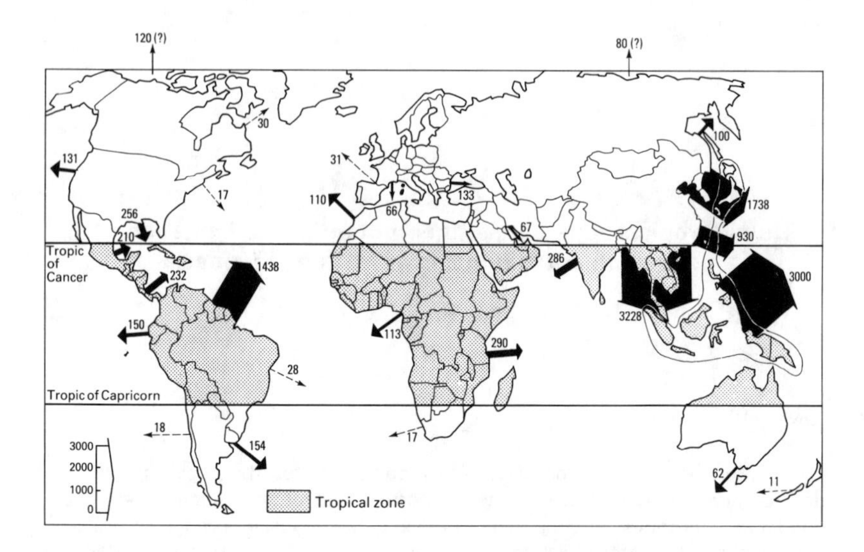

Fig. 1: Average annual input of suspended sediment from the drainage basins of the world (in millions of metric tons) into coastal waters (modified after Milliman and Meade, 1983 and Dyer, 1986).

impose limitations on development.

Sediment transport rates in rivers of the humid tropical zone are not, by definition, always high. The Congo (or Zaïre) river e.g. is a clear exception to the rule (Fig. 1). The heavily forested character of the catchment area, which is also relatively flat, limits denudation and sediment supply (Jansen, 1979).

In this paper attention is focused on river deltas in the Indonesian archipelago. Rivers and their associated deltas on the large islands of Java, Sumatra, Kalimantan, Sulawesi and Irian Jaya contribute significantly to the input of sediment into the coastal waters (Fig. 1). Especially on Java, soil losses and sediment supply are considerable. Some drainage basins of the largest rivers of Java lose an amount of sediment at a rate (River Solo: ca. 1200 $ton.km^{-2}.yr^{-1}$ or River Brantas: ca. 1650 $ton.km^{-2}.yr^{-1}$; Hoekstra, 1989a) which is only matched by e.g. the amount of sediment which is carried to the Yellow River (ca. 2500 $ton.km^{-2}.yr^{-1}$) or the Ganges-Brahmaputra system (1150-1500 $ton.km^{-2}.yr^{-1}$; data based on Coleman, 1982 and Jansen, 1979).

Case-studies from Java are taken to illustrate the impact of the climatic and environmental factors, to demonstrate the way in which local and man-induced factors and processes may interact with the environmental conditions and to analyse the management problems generally arising with respect to river deltas. The last part of the paper is essentially dealing with methods to improve the management of the deltaic systems, e.g. by changing the morphological and hydrodynamical characteristics of the deltaic regimes or by increasing the degree of planning and by avoiding improper land use.

Environmental conditions

The Indonesian island of Java is one of the most densely populated regions in the world. The greater part of this population is living in the northern coastal plain which was partly formed in Holocene times. A series of modern river deltas has developed along this northern coastline, where one can find the most favourable conditions for delta growth. Positive elements are the abundant supply of sediment from a great number of relatively large rivers, the presence of a shallow shelf sea (Java Sea with Sunda Shelf) and the low energy wave and tidal regime (Hoekstra, 1989a and 1993). Conditions on the south coast are markedly different: the rivers are smaller, the adjacent Indian Ocean is deep and huge swells limit the rate of deposition. Besides, breaking waves and associated undertow carry the sediment offshore. The shape of the northern deltas is invariably cuspate and digitate (= birdfoot-like; Tjia et al., 1968). In some cases, deltas appear to have a strongly deviating morphology.

The deltaic regions in the north though are commonly scarcely populated and the new deltas only seem to be of limited potential use for mankind. Flooding of coastal areas as a result of subsidence due to compaction and coastal erosion in relation to shifting distributaries are well-known environmental risks. In addition to these more familiar problems, the managers and developers of the recently built Javanese deltas also have to deal with more specific problems.

Java is located at latitude 7° South and is part of the humid tropics. From December until March the W-NW monsoon is dominant, causing abundant precipitation. The E-SE monsoon between May and September contains hardly any moisture; this period is the dry season. The mean annual precipitation varies from 1500 mm in the lower parts of the floodplains and coastal plain to more than 3000 mm higher up in the mountains. By far the largest part of the precipitation occurs during the wet season (Hoekstra, 1989a).

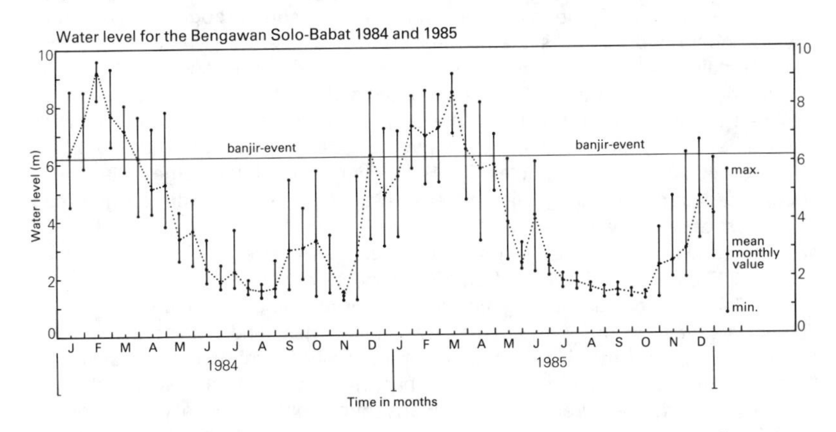

Fig. 2: Discharge distribution for a typical monsoonal river, the Solo river, East Java (data Proyek Bengawan Solo and Institute for Hydraulic Engineering, Bandung).

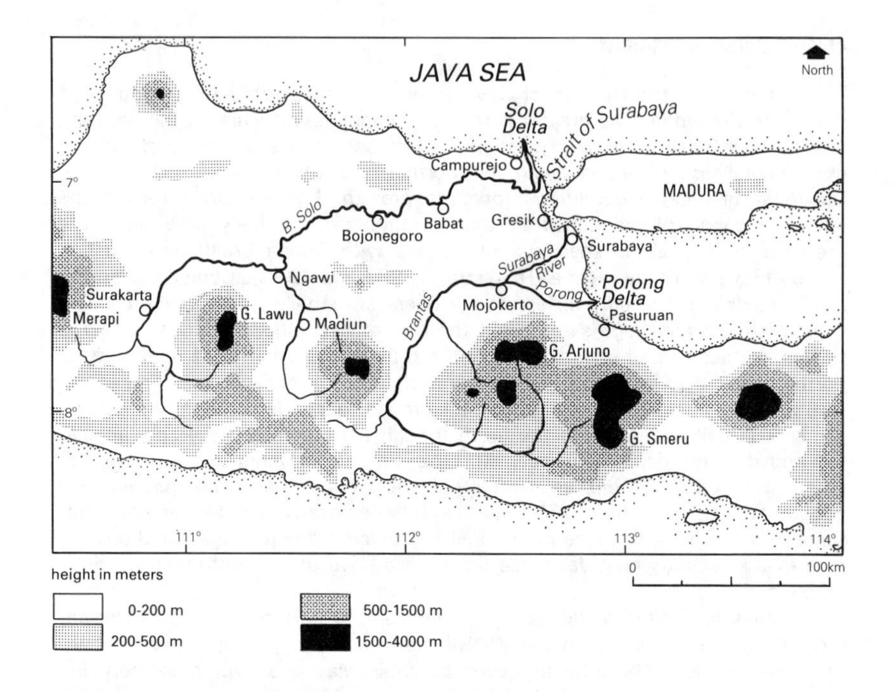

Fig. 3: Map of Central and East Java, illustrating the topographic setting of e.g. the drainage basins of Solo and Brantas.

The impact of the monsoonal climate is seen throughout the entire fluvial, deltaic and marine system. Most Javanese rivers are monsoonal rivers and the major part of river discharge takes place during the four or five months of the wet season (Fig. 2). It clearly means that there is a great contrast between the wet and dry season flow regime. During seasonal flash floods in the wet season - so-called banjirs - river discharges may increase to 2500 $m^3.s^{-1}$ or 4000 $m^3.s^{-1}$ (examples for Solo river, one of the largest Javanese rivers; see location map Fig. 3), whereas in the dry season the discharge drops to values of less than 20-30 $m^3.s^{-1}$. In extremely dry seasons, river flow is sometimes almost completely reduced to zero, as noticed e.g. in the Brantas river in 1972 (Fig. 3; MacDonald and Partners, 1977).

This monsoonal river regime has also far-reaching consequences for the deltaic systems. In the wet season, river discharges (and sediment load) are extremely high and the river deltas become entirely fluvial-dominated systems. River outflow commonly has a jet-like character. By contrast, river discharges in the dry season are very low and many river deltas can be considered as partially-mixed or well-mixed estuaries. The bi-directional transport of water in the river outlets of the deltas causes salt seawater to penetrate into the river channels. This process is often aided by the downstream configuration of these channels, their low gradients and the natural differences in density. In periods of extreme drought salinities in river deltas are often as high as $30^{0}/_{00}$.

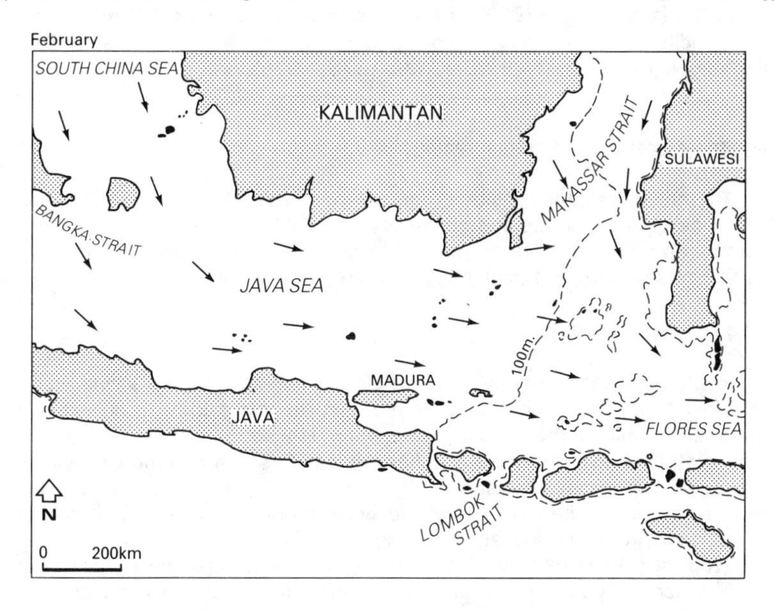

Fig. 4: Monsoon-induced seasonal flow patterns in the Java Sea during the wet season (February).

In the Solo river this salt intrusion not only affects the lower course of the river in the deltaic plain. The influence of brackish water (salinity $0\text{-}5^{0}/_{00}$) may be measured even as far as 80-85 km upstream from the river mouth (data Proyek Bengawan Solo-PBS; PBS-Monenco, 1984a). For rivers with a greater downstream slope, such as the Brantas/Porong in the east, this effect is less pronounced.

The fluvial-dominated character of many deltaic systems (for wet season conditions) is also due to a lack of a high energetic tidal regime. The north coast of Java is subject to a diurnal tide with a microtidal range (< 2 m). Tidal ranges vary from about 0.9 m at neap tide to ca. 2 m at spring tide. The associated tidal currents are generally only weak (Hoekstra, 1989a). The coastal flow patterns are actually dominated by a seasonal, wind-driven circulation. Monsoonal winds (again!) generate a seasonal flow pattern in the Indonesian archipelago. Along the northern coastline of Java, residual surface currents in the wet season (November-March) are directed eastwards and have a strength of 0.20 - 0.35 $m.s^{-1}$ (Fig. 4). The change in the monsoon (March-April) results in a western flow with maximum residual surface currents of about 0.25 $m.s^{-1}$ in July or August (data KNMI-Royal Dutch Meteorological Institute).

As mentioned before, wave activity on the north coast of Java is certainly limited. In the wet season deltas along the north coast are exposed to incoming waves from the NW. In 10% of the events, wave height exceeds 2 or 2.5 m but is commonly (90% of the observations) less than 1.5 m. For dry season conditions, the deltas are subject to low waves (95% less than 1.5 m) coming in obliquely from the E-NE (KNMI, 1975, 1976).

Management problems in deltaic regions

The environmental conditions for Javanese river deltas, in combination with a range of local factors and processes, inevitably lead to a series of major problems with concern to water quality and quantity, the excessive supply and deposition of sediment and conflicting or improper land use.

Water quality and quantity

In the dry season run-off is often almost neglectable and saline water penetrates into the deltaic channels. In many cases this salt intrusion extends far upstream. Most of the deltaic groundwater is also very saline and (fresh) drinking water - either from surface flow or aquifers - is commonly lacking. Human settlements can only survive if there is a regular supply of drinking water, e.g. by boat. The lack of high quality drinking water also is a limiting factor for agricultural and industrial activity.

Tropical soils within the deltas are negatively affected by the saline or brackish water. Only in the wet season, sufficient amounts of fresh water are available for irrigational purposes. Unfortunately, however, in that season river deltas become regularly flooded. The local population tries to avoid the risks of flooding by reinforcing the natural levees along the rivers and by building their villages on these levees or former beach ridges. The presence of the natural levees is actually a positive, natural response to the frequent overbank flooding.

Sediment supply and excessive deposition

The geological backbone of the island of Java is formed by a number of large volcanoes. Therefore the sediment load of the Javanese rivers is not only a result of the humid tropical climate but is also a reflection of the local geology with a high mountainous relief and a huge production of volcanic debris. The volcanic soils, in turn, are the basis for an intensive agriculture. According to Tjia et al. (1968) the amount of erosion and the rate of coastal accretion has increased significantly since 1910. They assume that large-scale deforestation due to the population boom, the Japanese occupation and the armed revolution was partly responsible for higher erosion rates. The input of sediment by the monsoonal rivers generally exceeds the transport capacity of coastal currents and waves. In the wet season river outflow processes are characterized by a jet-like flow. In these conditions, flow inertia is a major factor controlling delta dynamics. Quite often the large gradients in density between (fresh) river water and seawater give rise to the formation of buoyant jets and the development of buoyant plumes. In a seaward direction these plumes are subject to a lateral spreading and a vertical thinning (Hoekstra,

1989b). The associated gradual reduction in flow velocities with distance from the mouth results in a rapid or even exponential decrease in sediment load and high deposition rates.

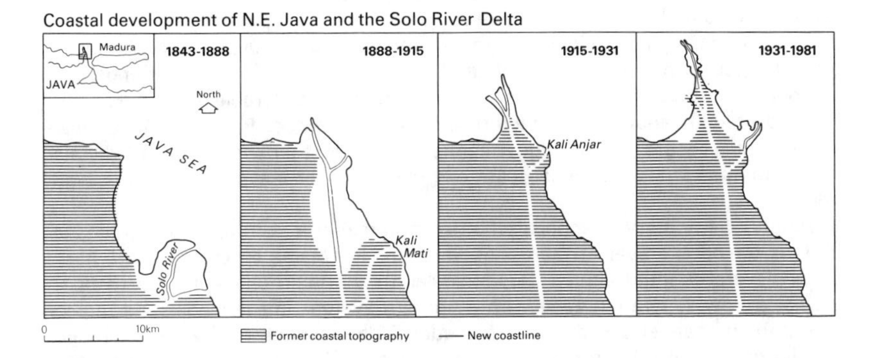

Fig. 5: Reconstruction of the coastal development of Northeast Java and the Solo delta since 1843.

It is essentially the combination of the presence of natural levees, more or less "channelizing" river outflow, the role of flow inertia and the rapid deposition of sediment in front of the river outlets which determine the growth pattern of many Javanese delta systems. The lateral dispersal and deposition of sediment is rather limited. As a result, the deltas or the individual branches within the deltas are often characterized by an elongate morphology.

An extreme example of a fluvial-dominated growth pattern is shown in Fig. 5. It illustrates the development of the elongate and rapidly prograding Solo delta in the last 150 years. At the beginning of the 20th century, the presently active Solo delta started to form and this delta has now reached a length of approximately 12 km (Fig. 5 and table 1). The delta is made up of one major straight channel. This shape though is rather exceptional and many other deltas, with more distributary channels, are commonly of the birdfoot-type. This is indicated by the growth figures presented in table 1. Many river branches of individual deltas are characterized by a rapid linear or longitudinal growth (table 1). The elongation-ratio, being the quotient of the surface area of the delta (or of a single branch, in km^2) and the seaward extension (= length, in km) of a major river branch, is generally small.

From a development point of view the presence of a highly indented coastline with multiple, extended distributaries, is a very inefficient morphology. The limited amount of surface area in comparison to the large extension of its coastline which has to be protected, makes development of such a delta not very beneficial.

The rapid longitudinal growth of the deltas also implies another problem: an accurate knowledge of the deltaic topography and hydrography is often lacking. For many coastal areas, old Dutch topographical maps from the mid fourties are the most recent maps (see e.g. the data presented in table 1;

the "latest" figures are regularly related to conditions in 1946).
The strong coastal progradation means that (former) fishing villages become isolated from the coast. Especially when shifting distributaries are involved - due to the almost continuous changes in hydraulic gradients - this will be a serious problem. The villages may have to be removed repeatedly.

Another major problem related to the excessive supply of sediment is the silting-up of navigable channels and ports. First of all, the rapid deposition in front of the river outlets creates mouth bars of considerable extend (Fig. 6). The waterdepth above these mouth bars is very limited: less than 1 m. Only small fishing boats like outrigger-canoes are able to operate in the deltaic environments. But even these boats can only enter or leave the deltas at high tide.
Secondly, distributary channels, whether man-made or with a natural origin - a bifurcation or a crevasse - are not very stable. Many of these distributary channels are oriented almost perpendicular to the main channel. Due to flow inertia, a flow-separation zone with a vortex develops in the inner bend region of the main channel and distributary outlet. The presence of a vortex stimulates the growth of a large point-bar which effectively reduces the cross-sectional area at the entrance of the channel. Consequently, these channels become increasingly blocked on both their landward side (by the point-bar) and at their seaward end (by the mouth bar). The distributary channels rapidly silt-up and are subject to a fast decline (Hoekstra, 1993).

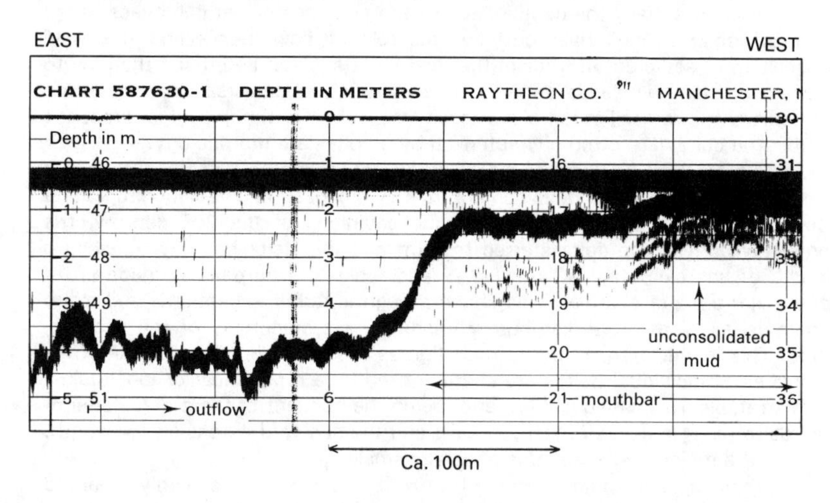

Fig. 6: Mouth bar in one of the western outlets of the River Solo.

Part of the river-borne sediment escapes from the deltaic environments, is seized by offshore coastal currents and is mainly transported longshore (compare Fig. 4; Hoekstra et al., 1989). In a downdrift direction (wet season: to the east) this causes a direct threat to the navigability of other channels and ports. For instance, the main fairway to the port of Surabaya (NE Java) and

Table 1: Javanese river deltas, growth rates and elongation-ratio; sources: Tjia et al. (1968), Hehanussa (1981) and Hoekstra (1989).

Delta system	Name of branch or location	Period	Linear growth L (m)	Linear growth rate ($m.yr^{-1}$)	Surface area A (km^2)	Areal growth rate ($km^2.yr^{-1}$)	Elongation ratio (A:L) (L in km)
Cipunegara		1865-1946			25.75	0.318	
	Pancer Kulon	--	4150	51			< 6.2
	Pancer Wetan	--	5500	68			< 4.7
	west	--	700	9			--
Cimanuk		1857-1946			51.10	0.574	
	Cimanuk	--	6400	72			< 8.0
	Anjar	--	9300	104			< 5.5
	Rambatan	--	1000	11			< 51.1
	Cemoro	--	2900	33			< 17.6
	Pancer Balok	1947-present	--	200	--	--	--
Cisanggarung	west	1864-1922	3520	61	10.88	0.188	3.1
	east	1922-1946	7680	320	22.25	0.927	2.9
Kali Pemali	west	1865-1946	2400	30	15.25	0.188	< 6.4
	east	1865-1946	4200	52			< 2.9
Kali Comal	west	1870-1946	--	--	9.24	0.122	--
	east	1870-1946	2080	27			< 4.4
Kali Bodri	--	1864-1946	1100	13	21.85	0.266	19.9
Kali Bujaran	--	1920-1946	1100	42	3.3	0.127	3.0
Bengawan Solo	--	1888-1984	12000	125	33.6	0.238	2.8
Kali Brantas	--	1886-1984	3200	33	47.7	0.486	14.9

the port itself do suffer from deposition. Already in the mid 19th century port authorities of Surabaya had to face this problem. In those days the problem was partly solved by diverting the Solo river (at about 1880) in a more northerly direction (Fig. 5). Nowadays with increasing amounts of trading vessels with heavy draught, the problem has to be tackled again. Large container vessels e.g. have to anchor in the Strait of Surabaya and to unload part of their cargo before they can moor to the quay.

The rapid deposition of sediment in the deltaic environments itself or in a downdrift direction also produces severe damage to vulnerable and productive tropical ecosystems like coral reefs. Reefs but also shoals of fish are often to be found in the vicinity of river mouths. The input of river water ensures marine organisms of a regular supply of nutrients. A high degree of turbidity and the deposition of sediment, however, limit the growth and development of reef-building organisms and eventually lead to a 'sudden death' of the reefs.

Present deltaic management and land use

The present coastal management of deltaic regions is quite often not in favour of future possibilities for development. The high deposition rates and the rapid growth of river deltas is partly enhanced by the presence of another tropical ecosystem: a natural mangrove vegetation, occupying the muddy tidal flats. The mangrove forests, fringing the deltas, effectively reduce the amount of coastal erosion due to currents and waves. Apart from this physical role, mangroves also have a major biological role in the tropical, marine ecosystem.

Nevertheless, in the early phases of (uncoordinated) development mangrove forests have to give way and are replaced by tambaks (= fish ponds) for aquaculture or salt pans. In addition, mangrove trees are cut by the first inhabitants to provide firewood and material for construction (fish-traps e.g.). Only a very small belt with mangrove trees is left along the coast as a rudimentary protection of the inland areas of the delta.

In general, one can say that human settlement results in a significant loss of mangrove forests. It may cause a number of effects. Firstly, the amount of coastal erosion may be enhanced, reducing the size of the deltas again and increasing the risks of flooding. Secondly, the mangrove forests act as a nursery for marine organisms, such as e.g. shrimps. Many marine organisms find - directly or indirectly - their food in the mangroves and mangrove forests form an essential part of the coastal food-chain. By cutting the mangroves, coastal water bodies become less productive (fisheries!). Paradoxically, the increased establishment of fish ponds to increase the productivity of aquaculture may even be - in the long run - disadvantageous for the same aquaculture.

Finally, mangrove forests are also favourable spots for mosquitoes. Mosquitoes are responsible for the spreading of typical tropical diseases like malaria. Malaria is endemic in many tropical countries and regions and the local population is, to some extent, immune for the disease. However, if one starts cutting the mangrove forests, the mosquito population will move and spread over the country-side, increasing the number of malaria victims. This can have catastrophic effects since the present medicines against malaria have already partly lost their effectiveness or have to be prescribed at high dosage.

In case mangrove forests are turned into salt pans, the high concentrations of salt and the presence of saline water, infiltrating into the soils, will destroy the physical and chemical structure of the soils. Therefore, any future use for agricultural purposes becomes almost impossible. Besides the presence of this improper land use, there is a great chance of conflicting land use as well. Examples are the development of industrial plants in areas with vulnerable ecological (and economical) activities such as fisheries.

Deltaic management and opportunities for development

The deltaic problems, discussed in the previous sections, are mainly related to natural factors and processes such as climate and local geology. It is clear that these factors are simply autonomous and cannot be affected by mankind. The best one can do is to adjust to these environmental conditions and to avoid any negative interaction with local conditions. In spite of these limitations, an improvement of deltaic management is always possible for certain aspects.

It is by no means the intention of the author to give a blueprint for development. Local conditions for Javanese deltas may differ considerably and emphasis in this paper is mainly given to natural processes. Socio-economic aspects of development are not taken into account; a cost-benefit analysis is also not included. For that reason, the practical applicability of the suggested measures for individual deltas should always be evaluated in a more detailed and comprehensive management study. The most profitable effects can probably be achieved by changing the characteristics of the fluvial and deltaic regimes. Three categories of measures are distinguished:

- sediment control methods in catchment areas and river basins;
- channel control and river training works inside the river deltas;
- increased planning activities at an early stage of development.

Sediment control methods

Measures in the catchment areas or river basins have to be taken with the main purpose of avoiding the large fluctuations in river discharge and reducing the huge (suspended) sediment load. Vegetative land treatment measures (such as reforestation) can be effective in reducing erosion by improving the protective cover on the soil surface and by increasing the infiltration rates.

Terracing is already a widespread method on Java for many centuries now. Reservoirs are commonly built to have sufficient amounts of water for irrigation. They can also be used successfully as a floodwater retarding structure to regulate river discharge in a downstream direction. In the dry season it may partly prevent the intrusion of saline water.

The reservoirs have to be extremely large to accept the overwhelming amounts of water during the seasonal flash floods in the wet season. However, a significant part of the sediment load, transported by the rivers is retained in the reservoirs (PBS-MONENCO, 1984b). The reservoirs are rapidly filled up with sediments and their effective capacity is seriously reduced. Therefore, debris basins should be built in conjunction with reservoirs. In addition, check dams and sediment traps are essential tools for a profitable use of the reservoirs. For many decades, reservoirs have been built and check dams are constructed in Central and North Java. But even nowadays, the impact of the river works is generally not large enough to have acceptable hydrodynamic conditions within the river deltas.

River training works inside the deltas

The main goal for introducing river training works inside the deltas is to create a better distribution of sediments. These training works have to stimulate the growth of a more cuspate or lobate type of delta instead of the frequently observed birdfoot-type of delta.

First of all, it is very important that at an early stage of development of the deltas no dykes are introduced or levees are reinforced. This allows a regular flooding of the deltaic systems and encourages deposition of sediment in the entire deltaic environment.

Secondly, in a later stage, distributary branches or outlets have to be created since they are essential for a lateral dispersal of sediment. It is very crucial that the bifurcations have a stable character. The development of a flow-separation zone with a vortex has to be avoided by making bifurcations with gradually curved channels. Curved distributary channels and outlets will be relatively stable if they have a steeper seaward slope than the main channel. In case of suspended load rivers the sediment load is commonly distributed over the two branches, according to the discharge distribution. For rivers with dominantly bedload transport, the presence of a curved channel with a helical flow pattern may give rise to a selective transport into mainly one river branch. Further downstream in the branch, river flow is generally not capable of transporting all the sediment. It results in silting-up of the river bed in the specific branch and the closing-off of this branch. Therefore, the ratio of bedload transport versus suspended load transport in a specific deltaic system has to play a major role in the decision about the configuration of the bifurcations.

A third problem, not easy to tackle, is the development of extensive mouth bars in many river outlets. A possible solution is the concentration of the outflowing effluents in relatively narrow channels. The reduction in cross-sectional area will lead to an increase in flow velocities, which causes an additional scour of the river bed in the river mouth. Jetties are probably too expensive for this purpose and their effect may be disadvantageous by stimulating the longitudinal growth again. But local techniques, such as shields of bamboo, oriented obliquely to the flow may be an adequate measure for partly controlling the flow and river bed.

A fourth problem, the downdrift transport of sediment from the deltas to adjacent coastal regions should be limited by directing the sediment output mainly in an "updrift" direction along the coast. In this way the delta acts as a barrier for the longshore (sediment) transport and the delta can intercept a significant part of the sediment. It will be in favour of deposition and delta growth on the updrift side of the deltas.

Planning activities

A number of suggested measures to improve the deltaic flow regime definitely requires the presence of a kind of management scheme already at an early stage of development. In general, one has to be careful with early settlements and land use. It may seriously limit the possibilities of development in a future situation. Especially mangrove forests have to be maintained in order to protect the deltas against erosion and to encourage continuous deposition. To some degree, tambaks or fish ponds should be allowed as long as they are no serious threat to the "coastal defence system". As a matter of fact, in many coastal regions of Java, tambaks probably are the best possible way to exploit the recently built deltas.

Conclusions

The management of river deltas in the humid tropical zone in general and on the Indonesian island of Java in particular is rather complicated due to the very specific climatic and environmental conditions. Especially on Java the highly variable discharge, the huge supply of sediment and the dominantly elongate or birdfoot-type morphology of the deltaic systems are the main, natural factors held responsible for the limited development. Actually, it is no surprise that most of the deltas are only scarcely populated.

Some improvement is expected if one is able to change or partly modify the fluvial and deltaic flow regimes. Some proposed measures will lead to a better control and an increased stability of deltaic river branches. In other cases though, the outcome of certain measures may be questionable and highly unpredictable due to the monsoonal flow regime. But even in the former case, a cost-benefit analysis will always be essential to evaluate the socio-economic value and effects of the river training works.

References

Coleman, J.M., 1982, Deltas. Processes of Deposition and Models for Exploration. International Human Resources Development Corporation, Boston, 124 pp.

Dyer, K.R., 1986, Coastal and Estuarine sediment dynamics. John Wiley, Chichester, 342 pp.

Hehanussa, P.E., 1981, Excursion to Cimanuk Delta. National Institute of Geology and Mining - Indonesian Institute of Sciences, 29 pp.

Hoekstra, P., 1989a, River outflow, depositional processes and coastal morphodynamics in a monsoon dominated deltaic environment, East Java, Indonesia. Thesis, Geografisch Instituut Rijksuniversiteit Utrecht, 215 pp.

Hoekstra, P., 1989b, Buoyant river plumes and mud deposition in a rapidly extending tropical delta. Proc. Int. Symp. on the results of the Indonesian-Dutch Snellius-II Expedition, Jakarta 1987 (LIPI-NRZ-UNESCO). Neth. J. of Sea Research, 23 (4): 517-527.

Hoekstra, P., 1993, Late Holocene development of a tide-induced elongate delta, the Solo delta, East Java. In: Woodroffe, C. (ed.): Evolution of coastal and lowland riverine plains in Southeast Asia and Northern Australia; Sedimentary Geology (in press).

Hoekstra, P., Nolting, R.F. and Van der Sloot. H.A., 1989, Supply and dispersion of water and suspended matter of the rivers Solo and Brantas into the coastal waters of East Java, Indonesia. Proc. Int. Symp. on the results of the Indonesian-Dutch Snellius-II Expedition, Jakarta 1987 (LIPI-NRZ-UNESCO). Neth. J. of Sea Research, 23 (4): 501-515.

Jansen, P.Ph., ed., 1979, Principles of River Engineering. Pitman London, 509 pp.

KNMI, 1975, Marine Climatological Summaries for the Mediterranean and the Southern Indian Ocean, 1966. De Bilt, The Netherlands 5: 180-200.

KNMI, 1976, Marine Climatological Summaries for the Mediterranean and the Southern Indian Ocean, 1967. De Bilt, The Netherlands 6: 183-206 and 7: 183-206.

MacDonald and partners, 1977, Lower Brantas Pollution Study. Government of Indonesia, Ministry of Public Works and Power, Directorate General of Water Resources Development, MacDonald, Cambridge: 1-53 with annex.

Milliman, J.D. and Meade, R.H., 1983, World-wide delivery of river sediment to the oceans, Journal of Geology 91: 1-21.

PBS-MONENCO, 1984a, Tidal effect study. In: Memorandum M-376. Lower Solo Project, Montreal Engineering Company Surakarta, Indonesia.

PBS-MONENCO, 1984b, Jipang Project Feasibility Study - Lower Solo River Development Study, Montreal Engineering Co. - Proyek Bengawan Solo, Surakarta, Indonesia, 3: 7-17.

Tjia, H.D., Sukendar Asikin and Atmadja, R.S., 1968, Coastal Accretion in Western Indonesia. Bulletin of National Institute of Geology and Mining, Bandung 1: 15-45.

The Fraser River delta, British Columbia: architecture, geological dynamics and human impact

J.L. Luternauer[1], J.J. Clague[1], J.A.M. Hunter[2], S.E. Pullan[2], M.C. Roberts[3], D.J. Woeller[4], R.A. Kostaschuk[5], T.F. Moslow[6], P.A. Monahan[7] and B.S. Hart[8]

Abstract

The Fraser River delta is the largest delta in western Canada. It is an important coastal ecosystem and an area of explosive urban and industrial growth lying within the most seismically active zone in Canada.

The Fraser delta has formed since the area was deglaciated about 13,000 years ago. Its lowest

---

[1]Research Scientist, Ph.D., P.Geo., Geological Survey of Canada, 4th Floor, 100 West Pender Street, Vancouver, B.C., Canada, V6B 1R8

[2]Research Scientist, Geological Survey of Canada, 601 Booth Street, Ottawa, Ontario, Canada, K1A 0E8

[3]Professor, Ph.D., P.Geo., Department of Geography, Simon Fraser University, Burnaby, B.C., Canada, V5T 1S6

[4]President, P.Eng., Conetec Investigations Ltd., Unit 3, 9113 Shaughnessy Street, Vancouver, B.C., Canada, V6P 6R9

[5]Associate Professor, Ph.D., Department of Geography, University of Guelph, Guelph, Ontario, Canada, N1G 2W1

[6]Associate Professor, Ph.D., Department of Geology, University of Alberta, Earth Sciences Building, Rm 1-10, Edmonton, Alberta, Canada, T6G 2E3

[7]Graduate Student, School of Earth and Ocean Sciences, University of Victoria, mailing address: 1024 Benvenuto Avenue, R.R. 1, Brentwood Bay, B.C., Canada, V0S 1A0

[8]Geomarine Consultant, Ph.D., c/o Geological Survey of Canada, Pacific Geoscience Centre, P.O. Box 6000, Sidney, B.C., Canada, V8L 4B2

stratigraphic unit consists of Holocene prodelta mud. South of the main distributary channel, these sediments are conformably overlain by a thick succession of sandy foreset beds. North of this channel, the foreset sequence consists primarily of silt with thick, localized sand bodies. The uppermost unit is a complex of sandy distributary channel deposits capped by mud and peat of intertidal, floodplain, and bog origin.

The main geological hazards are failures of the delta slope and earthquake-induced liquefaction. Deposition at the river mouth contributes to slope failures, including one of more than $1x10^6$ m^3 with a headscarp within 100 m of a staffed lighthouse. Evidence of liquefaction of a shallow subsurface sand unit, probably caused by one or more large prehistoric earthquakes, is present over much of the delta plain.

Previous discharge of treated sewage onto the tidal flats has led to the concentration of toxic metals and organic detritus. Causeways and other large engineering structures have altered sediment and water dispersal patterns and contributed to the erosion of some coastal habitats. Erosion of the tidal flats also may be aggravated by dredging of sand to maintain channels and provide construction material.

Introduction

The Fraser River delta (Fig. 1), just south of Vancouver, British Columbia, is the largest delta in western Canada. It is an important agricultural and waterfowl area and a vital link in the Fraser River salmon fishery; it is also an area of explosive urban and industrial growth and lies within the most seismically active zone in Canada (Milne et al., 1978). Fraser River distributary channels are trained and dredged to maintain navigable depths for ships. Pollutants are discharged into the estuary from industrial and sewage plants. Jetties and causeways interfere with sediment dispersal and water circulation on the wide tidal flats fringing the subaerial delta platform, and dredge spoil and construction waste are disposed of on the delta slope. The objective of this report is to present an overview of previous and current earth science studies of the delta that can help guide urban and industrial development and maintain ecological habitats.

Setting

The Fraser delta extends 15-23 km west and south from a narrow gap in the Pleistocene uplands east of Vancouver

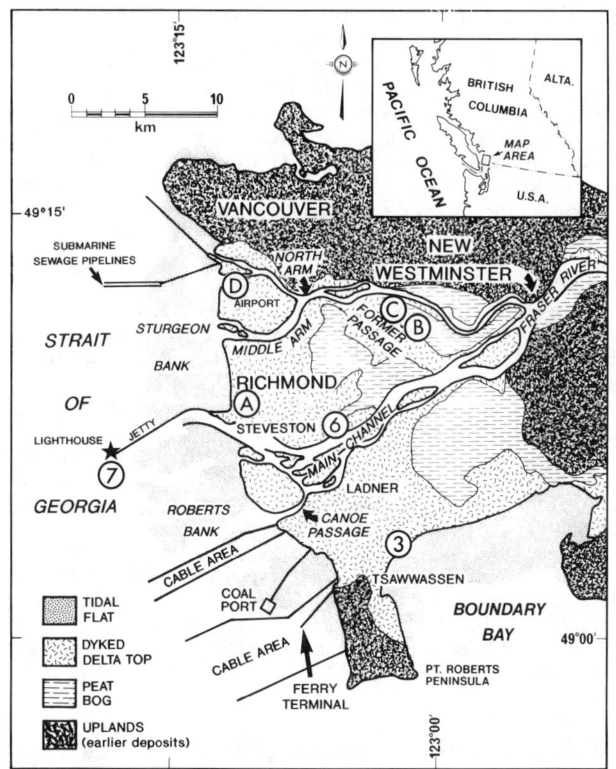

Figure 1. Location and setting of the Fraser River delta. Circled letters indicate locations where shear wave velocities plotted in Fig. 4 were measured. Circled numbers indicate locations of figures.

to meet the sea along a perimeter of about 40 km (Fig. 1). Twenty-seven kilometres of this perimeter, adjacent to the main distributary channels of Fraser River, face west onto the Strait of Georgia; the remainder face south onto Boundary Bay. These two sections are separated by Point Roberts peninsula, an upland and former island underlain by Pleistocene sediments. Very gently sloping tidal flats extend up to 9 km from the dyked edge of the delta to the subtidal foreslope. The western slope is inclined 1-23° (average 1.5°) towards the marine basin of the Strait of Georgia and terminates at about 300 m water depth, 5-10 km seaward of the tidal flats. The southern slope is ill-defined; it slopes more gently than the western foreslope and terminates in much shallower water (ca. 30 m).

Deposition on the Fraser delta is controlled by tidal and fluvial processes operating in a high-energy, semi-enclosed marine basin (Luternauer, 1980; Thomson, 1981).

The tidal range in the Strait of Georgia is relatively high (4-5 m), and all distributary channels are tidally influenced. Main Channel is affected by a salt-wedge intrusion, the position of which is controlled by river discharge and tidal height (Kostaschuk et al., 1992b). Sediment transport also depends on river and tidal conditions. At high tide the salt-wedge intrudes into the channel, coarse sediment is deposited at the landward edge of the wedge, and fine sediment continues to be carried seaward in the upper layer. Low tide flushes the salt-wedge from the channel, and bed material is resuspended and transported seaward.

Main Channel carries 80% of sediment supplied to the delta (Milliman, 1980), and sand constitutes about 35% of the 17.3 million tonnes of sediment discharged annually by Fraser River (McLean and Tassone, 1991). Prior to dyking, most of this sand was deposited on the tidal flats and slope adjacent to the distributary channels. Sediment and water discharge exhibit strong seasonal variations and are greatest during the late spring-early summer freshet (Thomson, 1981).

At the maximum of the last glaciation, ca. 14,000-15,000 BP, a grounded lobe of the Cordilleran Ice Sheet flowed south across the area now occupied by the Fraser delta. The delta formed subsequent to the disappearance of the ice sheet 11,000-13,000 BP (Johnston, 1921; Clague et al., 1983).

Surface sediments and present-day sedimentary environments

Present sedimentary environments of the Fraser delta include bogs, the floodplain, river channels, tidal flats and the foreslope (Clague et al., 1983) (Figs. 1, 2). Large domed peat bogs cover more than 80 km^2 of the eastern delta plain (Johnston, 1921; Hebda, 1977). They have developed on a poorly drained substrate close to, and locally below, low tide level. The continuity of the large peat bogs on the eastern Fraser delta attests to the stability of distributary channels there during late Holocene time. At only one place is there a gap in the peat deposits not now occupied by an active distributary channel (Fig. 1). The stability of channels on this part of the delta contrasts with the instability on the tidal flats, where there have been marked shifts in channel positions historically (Luternauer and Finn, 1983).

Surface sediments underlying much of the dyked portion of the delta plain are sandy to clayey silt deposited in overbank and uppermost intertidal environments. These sediments accumulated during spring floods and at times of very high tides. Distributary

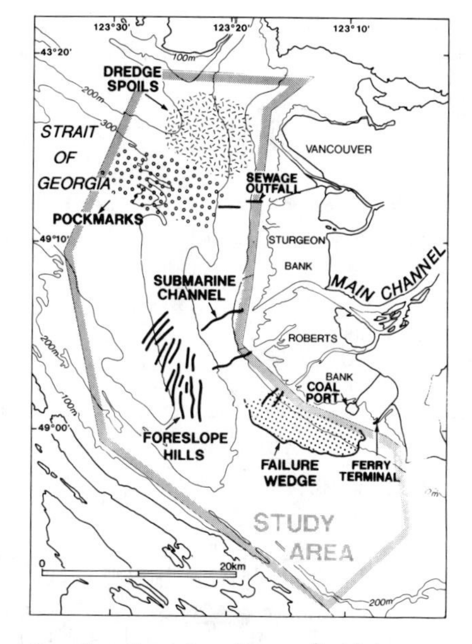

Figure 2. Surficial features of the Fraser delta slope and adjacent sea floor. (After Hart, 1992; Hart et al., 1992a,b; Hart and Hamilton, 1993).

channels are incised into these and underlying sediments to depths as great as 22 m below sea level, are floored by sand and, locally, gravel, and include sandy islands (Johnston, 1921; Mathews and Shepard, 1962).

Tidal flats are mantled mainly by flat-lying, fine to medium sand. A discontinuous fringe of marsh, underlain by muddy sediments, marks the landward edge of this zone. Sand swells with heights of 0.5 m and wavelengths of 50-100 m are common on the tidal flats of the western delta front and result from the movement of sediment by wind-generated waves and currents (Luternauer, 1980).

Foreslope sediments consist of fine sand and sandy to clayey silt. The coarsest sediments occur on the upper slope near the mouths of active distributary channels and over most of the slope south of Main Channel. Sand may be widespread in the latter area because this part of the delta presently is sediment-starved and an area of nondeposition or erosion (Luternauer, 1980). West and north of Main Channel, slope sediments gradually fine from sandy silt to clayey silt with increasing water depth and distance from the primary sediment source. The slope is cut by sea valleys, formed and maintained by slumps, turbidity currents and debris flows (Mathews and Shepard, 1962; Luternauer, 1980; Atkins and Luternauer, 1991; Kostaschuk et al., 1989, 1992a; Moslow et al., 1991, 1993; Hart et al., 1992a,b; McKenna et al., 1992; Evoy et al., 1993).

Stratigraphy and evolution

The subsurface of the Fraser delta plain and tidal flats is known primarily from boreholes drilled by the Geological Survey of Canada (GSC) and Simon Fraser

University (SFU) (Roberts et al., 1985; Williams and Roberts, 1989; Clague et al., 1991) and from engineering (especially cone penetration) test holes (Monahan et al., 1993). High-resolution, reflection seismic data acquired by the GSC and SFU are available for the southernmost part of the delta (Jol and Roberts, 1988, 1992; Pullan et al., 1989; Clague et al., 1991; Roberts et al., 1992). Elsewhere, the high gas content of the sediment obscures seismic returns.

Fraser delta sediments have a maximum thickness in excess of 200 m (Johnson, 1921) and unconformably overlie late Pleistocene till and stratified drift. The deltaic succession can be broadly divided into topset and foreset deposits (Fig. 3), and bottomset deposits. Topset sediments thin and climb to the west, reflecting seaward progradation of the delta under a regime of rising sea level during middle to late Holocene time (Clague et al., 1983; Williams and Roberts, 1989).

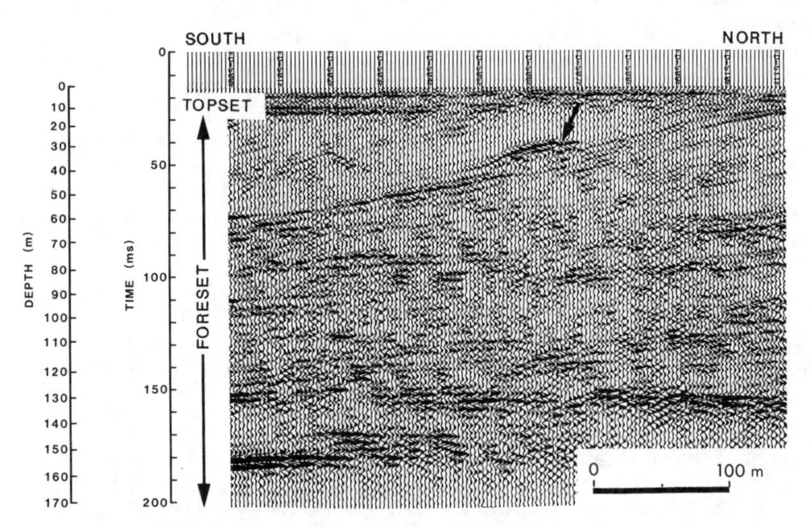

Figure 3. High resolution seismic profile from the southern part of the delta, showing flat-lying topset and gently dipping foreset beds. Small arrow points to buried failure deposit.

The lower part of the topset sequence is a 10-20 m thick sand unit interpreted to be a distributary channel complex (Monahan et al., 1993). This unit commonly is overlain by burrowed sand and silt interpreted to be an intertidal deposit (Williams and Roberts, 1989; Monahan et al., 1993). Organic-rich silt deposited in a floodplain environment overlies the intertidal sediments and, locally, the distributary channel deposits. Intertidal and floodplain sediments are up to 15 m thick on the eastern part of the delta, where they are as old as 8000 BP. In contrast they are only 5 m thick at the western margin of the dyked delta plain (Clague et al., 1983; Williams and Roberts, 1989).

The topset sequence is underlain by foreset beds deposited on the ancestral Fraser delta slope. North of Main Channel, the foreset sequence consists mainly of silt. Sand bodies up to 30 m thick in the upper part of this silty sequence may have been deposited adjacent to the mouths of distributary channels. Sand beds interbedded with silt deeper in the sequence may be gravity flow deposits that bypassed the upper slope. South of Main Channel, the foreset sequence consists mainly of sand beds dipping, on average, 2° seaward (Fig. 3).

The foreset sequence conformably overlies bottomset silt and clay (Clague et al., 1991). These sediments are similar to those that are presently accumulating in the deep basins of the Strait of Georgia west and northwest of the Fraser delta (Pharo and Barnes, 1976).

Airgun seismic profiles collected offshore show that deltaic deposits west of the tidal flats locally are thicker than 200 m, although they are absent over parts of some submarine ridges cored by bedrock and Pleistocene deposits. High resolution seismic profiles demonstrate that much of the slope, especially north of Main Channel, is underlain by undisturbed sediments; the uppermost 5 m of these sediments contain interstitial gas (Hart et al., 1992b; Hart and Hamilton, 1993).

Geohazards

The main geohazards are associated with earthquakes and failures of the delta slope. The susceptibility of sediments to liquefaction during earthquakes can be determined empirically using normalized shear wave velocity, V_{s1} (Robertson et al., 1992). Sediments with normalized shear wave velocities less than about 165 m/s are susceptible to liquefaction during earthquake shaking (Fig. 4). Ground motion amplification on the Fraser delta due to earthquake shaking has been estimated using shear wave velocity data from more than 70 sites (Hunter et al., 1992). Shear wave velocities at depths of less than 80 m are less than 300 m/s whereas those in underlying Pleistocene deposits are as high as 700 m/s. The average velocity at each of the Fraser delta sites has been compared to a reference bedrock velocity of 1500 m/s to obtain a relative amplification value. Values exceed 2.5 over most of the delta (Fig. 5); exceptions are areas where high-velocity Pleistocene sediments approach the surface. Shear-wave velocity data also have been used in one-dimensional models, such as SHAKE (Schnabel et al., 1972), to obtain ground-motion resonant peaks at various earthquake frequencies (Byrne and Anderson, 1987, 1991; Bazet and McCammon, 1986; Finn and Nichols, 1988).

Sand dykes and sills, which are common in near-surface sediments over much of the Fraser delta, provide evidence for earthquake-induced liquefaction (Clague et al., 1992) (Fig. 6). At a few sites evidence has been found for venting of sand onto a subaerial or intertidal surface. All observed sand dykes, sills, and boils are younger than 3500 BP, and at least some are younger than 2400 BP.

Failures on the delta slope could damage or destroy a sewage outfall and submarine transmission cables that supply electricity and telephone service to Vancouver Island. Such failure also could impact structures at the edge of the tidal flats, including a ferry terminal and Canada's largest coal export facility (Figs. 1,2). A large failure might also generate a tsunami in the Strait of Georgia (Hamilton and Wigen, 1987).

Failures have occurred on several scales both historically and in the past (Fig. 3). Deposition at the river mouth contributes to failures, including one larger than $1x10^6$ m^3 that heads within 100 m of a staffed lighthouse (Atkins and Luternauer, 1991; McKenna et al., 1992) (Fig. 7). Two large, submarine failure complexes have been identified at the base of the Fraser delta slope. One is the Foreslope Hills (Fig. 2), an area of curvilinear ridges and troughs up to 20 m high, covering an area of at least 60 km^2 (Tiffin et al., 1971). The ridges consist of discrete blocks of prodelta and foreslope sediments bounded by faults that dip seaward. The deformation may be the result of a complex type of failure involving lateral spreading and rotational slumping of foreslope sediments above clayey prodelta deposits (Hart et al., 1992b). A second submarine failure complex underlies at least 40 km^2 of the seafloor on the Roberts Bank slope (Fig. 2). Airgun seismic records show that this complex comprises a wedge of chaotic and wavy reflectors, locally over 75 m thick (Hart et al., 1992b). The causes of these failures are unknown, although earthquakes, rapid sedimentation, and interstitial gas may all have contributed.

Human impact

Development associated with the growth of Vancouver and its satellite communities has adversely affected the Fraser delta. Pollutants discharged into the Fraser River

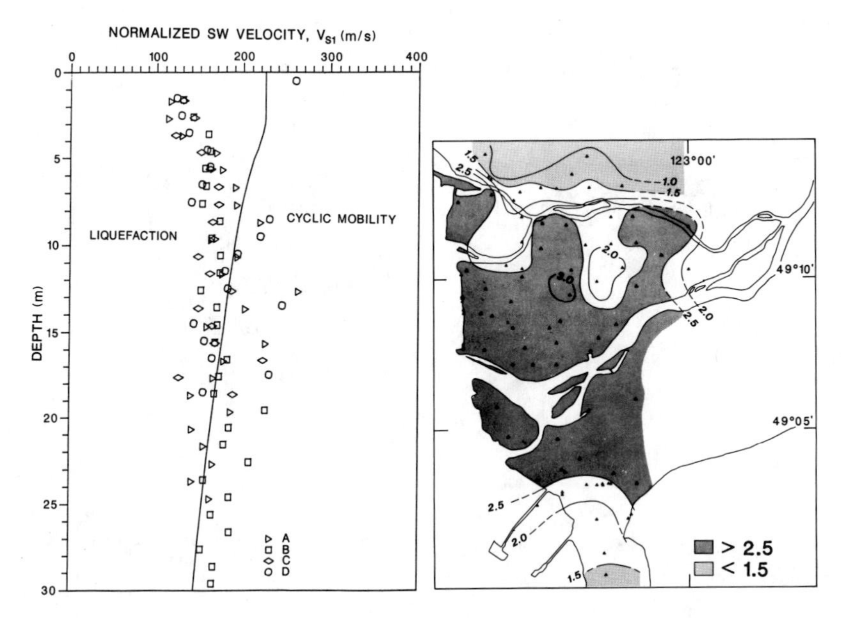

Figure 4. Normalized shear wave velocities at four sites in Richmond (see Fig. 1 for locations). The liquefaction threshold curve (for an earthquake of magnitude 7.5) has been calculated using the cyclic stress ratio-velocity relationship of Robertson (1990), and the proposed cyclic stress ratio depth curve for a peak ground acceleration of 0.3 g developed by Byrne and Anderson (1991).

Figure 5. Ground motion amplification on the Fraser delta relative to bedrock. Triangular symbols indicate data sites.

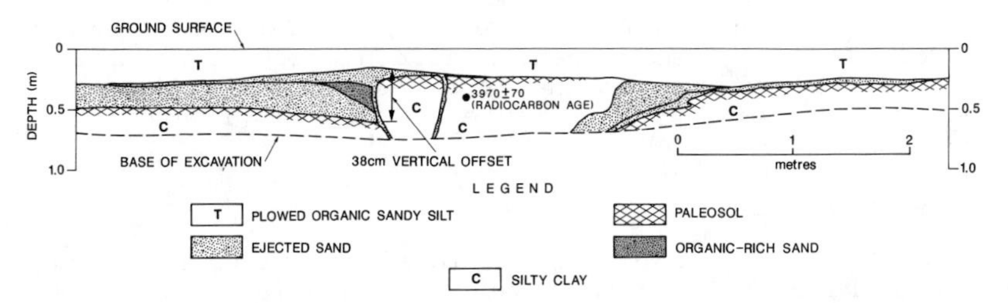

Figure 6. Profile of sand dykes and vented sand exposed in the shallow wall of an excavation on the central Fraser delta. The vented sand directly overlies a brown paleosol. It displays stratification indicative of emplacement during either multiple events or multiple pulses within a single event. Note the vertical displacement of the paleosol and the subjacent mud along the two main feeder dykes. (From Clague et al., 1992).

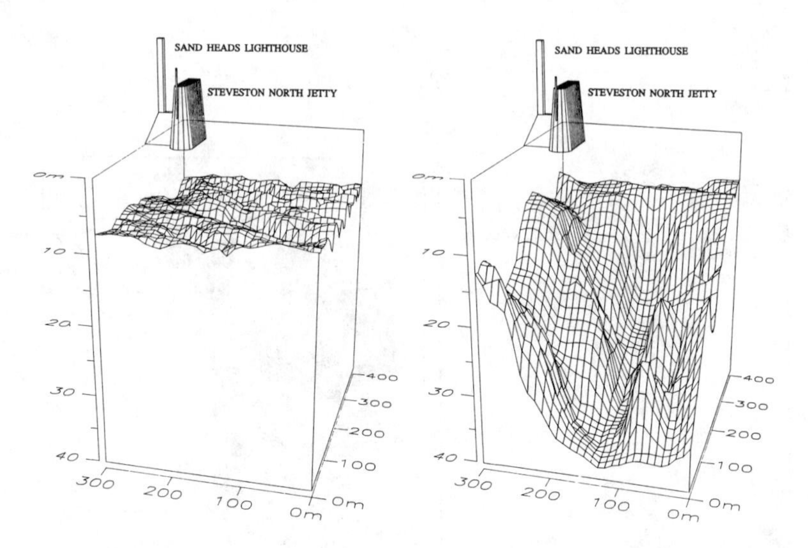

Figure 7. Perspective computer-generated images of the mouth of Main Channel before (June 27, 1985) and after (July 11, 1985) a large failure. Vertical exaggeration 10x, 0m = mean low low water. (computer drafting by R.J. Atkins).

include antisapstain chemicals, polycyclic aromatic hydrocarbons, and dioxins/furans (Standing Committee on the Fraser River Estuary Water Quality Plan, 1990). These toxins may be concentrated at a turbidity maximum identified at the upriver end of the salt wedge (Kostaschuk et al., 1992b). Discharge of treated sewage onto the tidal flats has led to the concentration of toxic metals and organic detritus (Birtwell et al., 1983). Causeways and other large engineering structures have altered sediment and water dispersal patterns (Levings, 1980) and contributed to the erosion of some coastal habitats (Duggan and Luternauer, 1985). Volume of sand dredged for river maintenance and construction material exceeds that supplied to the estuary by the river (McLean and Tassone, 1991). This may exacerbate erosion at the delta front.

Acknowledgments

J.V. Barrie critically reviewed the manuscript and offered helpful suggestions for improving it. Very many thanks to Bev Vanlier for typing and formatting it. This manuscript is Geological Survey of Canada Contribution 46492.

References

1. Atkins, R.J. and Luternauer, J.L., "Re-evaluation of slope failure volume estimates involving 3D computer graphics for the Fraser River delta front, British Columbia"; in Current Research, Part E; Geological Survey of Canada, Paper 91-1E, pp. 59-65, 1991.

2. Bazet, D.J. and McCammon, N.R., "Foundations of the Annacis cable-stayed bridge"; Canadian Geotechnical Journal, Vol. 23, pp. 458-471, 1986.

3. Birtwell, I.K., Greer, G.L., Nassichuk, M.D., and Rogers, I.H., "Studies on the impact of municipal sewage discharges onto an intertidal area within the Fraser River estuary, British Columbia"; Canadian Technical Report, Fisheries and Aquatics Sciences, No. 1170, 1983.

4. Byrne, P.H. and Anderson, D.L., "Earthquake design in Richmond, B.C., Version II"; Soils Mechanics Series, No. 109, Department of Civil Engineering, University of British Columbia, Vancouver, 1987.

5. __________, "Earthquake design in the Fraser Delta"; Task Force Report (Part 1), prepared for the City of Richmond, British Columbia, 1991.

6. Clague, J.J., Luternauer, J.L., and Hebda, R.J., "Sedimentary environments and postglacial history of the Fraser River delta and lower Fraser Valley, British Columbia"; Canadian Journal of Earth Sciences, Vol. 20, pp. 1314-1320, 1983.

7. Clague, J.J., Luternauer, J.L., Pullan, S.E., and Hunter, J.A., "Postglacial deltaic sediments, southern Fraser River delta, British Columbia"; Canadian Journal of Earth Sciences, Vol. 28, pp. 1386-1393, 1991.

8. Clague, J.J., Naesgaard, E., and Sy, A., "Liquefaction features on the Fraser delta: evidence for prehistoric earthquakes?"; Canadian Journal of Earth Sciences, Vol. 29, pp. 1734-1745, 1992.

9. Duggan, D.M. and Luternauer, J.L., "Development-induced tidal flat erosion, Fraser River delta, B.C."; in Current Research, Part A; Geological Survey of Canada, Paper 85-1A, pp. 317-326, 1985.

10. Evoy, R.W., Moslow, T.F., and Luternauer, J.L., "Sedimentology and lithofacies associations of the Fraser River delta foreslope, British Columbia"; in

Current Research, Part A; Geological Survey of Canada, Paper 93-1A, pp. 273-280, 1993.

11. Finn, W.D.L. and Nichols, A.M., "Seismic response of long-period sites: lessons from the September 19, 1985 Mexican earthquake"; Canadian Geotechnical Journal, Vol. 25, pp. 128-137, 1988.

12. Hamilton, T.S. and Wigen, S.O., "The foreslope hills of the Fraser delta: implications for tsunamis in Georgia Strait"; Science of Tsunami Hazards, Vol. 15, pp. 15-33, 1987.

13. Hart, B.S., "Side-scan sonar observations of Point Grey dump site, Strait of Georgia, British Columbia"; in Current Research, Part A; Geological Survey of Canada, Paper 92-1A, pp. 55-61, 1992.

14. Hart, B.S. and Hamilton, T.S., "High resolution acoustic mapping of shallow gas in unconsolidated sediments beneath the Strait of Georgia, British Columbia"; Geo-Marine Letters, 1993, in press.

15. Hart, B.S., Prior, D.B., Barrie, J.V., Currie, R.A., and Luternauer, J.L., "A river mouth submarine landslide and channel complex, Fraser delta, Canada"; Sedimentary Geology, Vol. 81, pp. 73-87, 1992a.

16. Hart, B.S., Prior, D.B., Hamilton, T.S., Barrie, J.V., and Currie, R.G., "Patterns and styles of sedimentation, erosion and failure, Fraser delta slope, British Columbia"; in Geotechnique and Natural Hazards [Proceedings of First Canadian Symposium on Geotechnique and Natural Hazards]; Canadian Geotechnical Society, Vancouver, May 1992, pp. 365-372, 1992b.

17. Hebda, R.J., "The paleoecology of a raised bog and associated deltaic sediments of the Fraser River delta"; Ph.D thesis, University of British Columbia, Vancouver, 1977.

18. Hunter, J.A., Luternauer, J.L., Neave, K.G., Pullan, S.E., Good, R.L., Burns, R.A., and Douma, M., "Shallow shear wave velocity-depth data in the Fraser River delta from surface refraction measurements, 1989, 1990, 1991"; Geological Survey of Canada, Open File 2504, 1992.

19. Johnston, W.A., "Sedimentation of the Fraser River Delta"; Geological Survey of Canada, Memoir 125, 1921.

20. Jol, H.M. and Roberts, M.C., "The seismic facies of a delta onlapping an offshore island: Fraser River Delta, British Columbia"; in Sequences, Stratigraphy, Sedimentology: Surface and Subsurface, (ed.) D.P. James and D.A. Leckie; Canadian Society of Petroleum Geologists, Memoir 15, pp. 137-142, 1988.

21. ________, "The seismic facies of a tidally influenced Holocene delta: Boundary Bay, Fraser River delta, B.C."; Sedimentary Geology, Vol. 77, pp. 173-183, 1992.

22. Kostaschuk, R.A., Stephan, B.A., and Luternauer, J.L., "Sediment dynamics and implications for submarine landslides at the mouth of the Fraser River, British Columbia"; in Current Research, Part E; Geological Survey of Canada, Paper 89-1E, pp. 207-212, 1989.

23. Kostaschuk, R.A., Luternauer, J.L., McKenna, G.T., and Moslow, T.F., "Sediment transport in a submarine channel system: Fraser River delta, Canada"; Journal of Sedimentary Petrology, Vol. 62, pp. 273-282, 1992a.

24. Kostaschuk, R.A., Church, M.A., and Luternauer, J.L., "Sediment transport over salt-wedge intrusions: Fraser River estuary, Canada"; Sedimentology, Vol. 39, pp. 305-317, 1992b.

25. Levings, C.D., "Consequences of training walls and jetties for aquatic habitats at two British Columbia estuaries"; Coastal Engineering, Vol. 4, pp. 111-136, 1980.

26. Luternauer, J.L., "Genesis of morphologic features on the western delta front of the Fraser River, British Columbia -- status of knowledge"; in The Coastline of Canada, Littoral Processes and Shore Morphology, (ed.) S.B. McCann; Geological Survey of Canada, Paper 80-10, pp. 381-396, 1980.

27. Luternauer, J.L. and Finn, W.D.L., "Stability of the Fraser River delta front"; Canadian Geotechnical Journal, Vol. 20, pp. 606-613, 1983.

28. Mathews, W.H. and Shepard, F.P., "Sedimentation of the Fraser River delta"; Bulletin of the American Association of Petroleum Geologists, Vol. 46, pp. 1416-1463, 1962.

29. McKenna, G.T., Luternauer, J.L., and Kostaschuk,

R.A., "Large-scale mass-wasting events on the Fraser River delta front near Sand Heads, British Columbia"; Canadian Geotechnical Journal, Vol. 29, pp. 141-156, 1992.

30. McLean, D.G. and Tassone, B.L., "A sediment budget of the lower Fraser River"; Proceedings, 5th Federal Interagency Sedimentation Conference, Las Vegas, Nevada, 1991.

31. Milliman, J.D., "Sedimentation in the Fraser River and its estuary, southwestern British Columbia"; Estuarine and Coastal Marine Science, Vol. 10, pp. 609-633, 1980.

32. Milne, W.G., Rogers, G.C., Riddihough, R.P., McMechan, G.A., and Hyndman, R.D., "Seismicity of western Canada"; Canadian Journal of Earth Sciences, Vol. 15, pp. 1170-1193, 1978.

33. Monahan, P.A., Luternauer, J.L., and Barrie, J.V., "A delta topset "sheet" sand and modern sedimentary processes in the Fraser River delta, British Columbia"; in Current Research, Part A; Geological Survey of Canada, Paper 93-1A, pp. 263-272, 1993.

34. Moslow, T.F., Luternauer, J.L., and Kostaschuk, R.A., "Patterns and rates of sedimentation on the Fraser River delta slope, British Columbia"; in Current Research, Part E; Geological Survey of Canada, Paper 91-1E, pp. 141-145, 1991.

35. Moslow, T.F., Evoy, R.W., and Luternauer, J.L., "Implications of ^{137}Cs sediment profiles for sediment accumulation rates on the Fraser River delta foreslope, British Columbia"; in Current Research, Part A; Geological Survey of Canada, Paper 93-1A, pp. 255-261, 1993.

36. Pharo, C.H. and Barnes, W.C., "Distribution of surficial sediments of the central and southern Strait of Georgia, British Columbia"; Canadian Journal of Earth Sciences, Vol. 13, pp. 684-696, 1976.

37. Pullan, S.E., Jol, H.M., Gagne, R.M., and Hunter, J.A., "Compilation of high resolution 'optimum offset' shallow seismic reflection profiles from the southern Fraser River delta, British Columbia"; Geological Survey of Canada, Open File 1992, 1989.

38. Roberts, M.C., Williams, H.F.L., Luternauer, J.L., and Cameron, B.E.B., "Sedimentary framework of the

Fraser River Delta, British Columbia: preliminary field and laboratory results"; in Current Research, Part A; Geological Survey of Canada, Paper 85-1A, pp. 717-722, 1985.

39. Roberts, M.C., Pullan, S.E., and Hunter, J.A., "Applications of land-based high resolution seismic reflection analysis to Quaternary and geomorphic research"; Quaternary Science Reviews, Vol. 11, pp. 557-568, 1992.

40. Robertson, P.K., "Seismic cone penetration testing for evaluating liquefaction potential"; Proceedings, Symposium on Recent Advances in Earthquake Design Using Laboratory and In Situ Tests, ConeTec Investigations Ltd., Vancouver, B.C., 1990.

41. Robertson, P.K., Woeller, D.J., Kokan, M., Hunter, J., and Luternauer, J.L., "Seismic techniques to evaluate liquefaction potential"; Proceedings, 45th Canadian Geotechnical Conference, Toronto, pp. 5-1 to 5-9, 1992.

42. Schnabel, P.B., Lysmer, J., and Seed, H.B., "SHAKE: a computer program for earthquake response analysis of horizontally layered sites"; Earthquake Engineering Research Centre, University of California, Berkeley, Report EERC 72-12, 1972.

43. Standing Committee on the Fraser River Estuary Water Quality Plan, "Status report on water quality in the Fraser River Estuary"; Fraser River Estuary Management Program, New Westminster, B.C., 1990.

44. Thomson, R.E., "Oceanography of the British Columbia coast"; Canadian Special Publication of Fisheries and Aquatic Sciences, No. 56, 1981.

45. Tiffin, D.L., Murray, J.W., Mayers, I.T., and Garrison, R.E., "Structure and origin of Foreslope Hills, Fraser Delta, British Columbia"; Bulletin of Canadian Petroleum Geology, Vol. 19, pp. 589-600, 1971.

46. Williams, H.F.L. and Roberts, M.C., "Holocene sea-level change and delta growth: Fraser River Delta, British Columbia"; Canadian Journal of Earth Sciences, Vol. 26, pp. 1657-1666, 1989.

Environmental Change in Southern Part of
Mekong River Delta and Problems of Territorial
Rational Use

Le Duc An[1]
Pham Trung Luong[1]

Abstract

Mekong River Delta (MRD) is a largest plain in South-East Asia. Its southern part as a peninsula is a recently forming plain with low surface and it has been made up by muddy silt.

Last time, due to human activities and natural processes the ecological environment of this area is strongly changed.

A study for the environmental impact assessment due to shrimp production in the Camau bay based on the system and interdisciplinary approach combining with remote sensing is carried out.

Introduction

The Mekong River Delta (MRD) is a largest plain in South - East Asia, which pays an important role in social-economical development of the country. In the MRD, there is a typical tropical mangrove fo - rest. Some places of the forest are of origin, so it is of great interest not only for managers but also for scientists.

Last time, due to intensive human activities and natural processes the ecological environment of MRD is strongly changed, that influences not only on country's social-economical development, but also on environment. By above mentioned, a project on MRD's environmental changes and human impact assessment is required. The resent study is one of those that have been carried out last time in this aspact in the Vietnam.

A Study Area

The Mekong river plain : The Mekong river with 4500 Km in long

- - - - - - - - - - - - - - - - - -

[1]Scientists, National Centre for Scientific Research of Vietnam, Nghia do, Tu liem, Ha noi, Vietnam.

and total basin area of about 800,000 Km^2,yearly carries average amount of 500 Km^3 of water and 70 millions tons of sediment. A large plain with total area of about 90,000 Km^2 has been formed in area where the river inputs to the sea. The MRD with less area (about 60,000 Km^2) has been formed in Miocene as a result of extention of depression Paleogene on eastern continental self /1/. The plain is limited in the north-east by young basaltic plateau (Q_I-Q_{II}) and its fundament was formed mainly by Mesozoic rocks, and in some where, by Paleozoic rocks. A simple synclinal with axis depth of 200-2200m has been formed by deposits cover Kainozoic (Fig.1). These deposits have ages of Neogence and Quaternary and their common particular is that coarse sand lays under and fine sand is distributed on the surfaceof investigated profile. The deposits with fine sand consit mixed layers of continental and marine origins with the main role of the last one. There were 3 times of transgression in middle Miocene, in Pliocene and in Quaternary. In Quaternary there were 3 transgression in the middle Pleistocene, in the end of the late Pleistocene and in the early-middle Holocene.

The horizontal surface with absolute high from 2-3m to 5m is one of particulars of MRD's plain. The fluvial terraces surrounding MRD have layering form of relief with 4 levels of high : 10-15m, 40-50m, 60-70m, and 90-100m.

The hydrological regime of MRD is devided to 2 seasons :pluvial season with duration of 5 months covers 75% yearly water amount and dry season of 7 months. In the pluvial season more than 2/3 MRD's area is flooded.

Camau peninsula : The southern part of MRD with total area of about 10,000 Km^2 is considered as a peninsula called Camau. All surface deposits have been already formed in period from Holocene up to present. The depth of fundamental bedrock in this area is changed from 400m in the west to 2200m in estuary of river Hau considered as the eastern point of the Camau peninsula.

In Nam can district, the Miocene deposit covered granite rock with thickness of 35m is distributed at the lower layers. This deposit consits kaolin clay and coarse sand of land facies. The continued is Pliocene deposit with thickness of 115m,which consits sand,fine sand, silt clay of shallow sea facies. The Quaternary deposit of sea origin with thickness of 250m consits grains with different size from coarse (sand,fine sand) to fine (clay,silt clay). In this area the attending of Holocene deposit of sea origin having a thickness to 40m shows that the sinking process is carrying out more intensive than before. The obtained analysis results of gained samples from low drillholes prove that from 5m depth to the surface, the accumu - lative environment is developed with tendency of decreasing salt strength, and evolved from low marine regime to marsh-mangrove one.The analysis results of C^{14} show that average accumulation speed from late Pleistocene up to present is about 3.5 mm/year (vegetation vestiges with age C^{14} of 38,100 $\pm$ 1200 years have been formed at the depth of 136m). In opposition, the average accumulation speed of su-

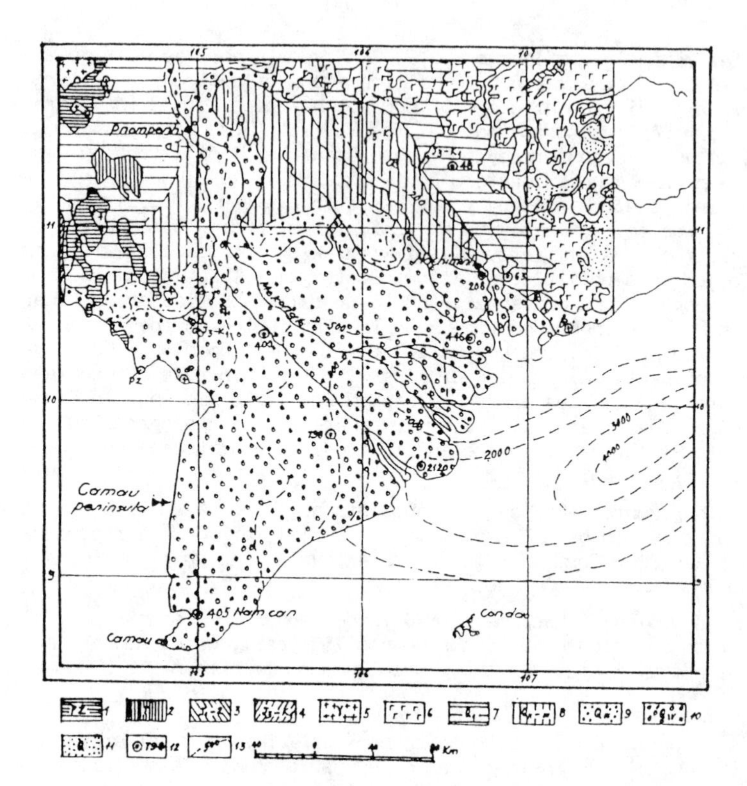

Fig 1. Geological sketch map of Mekông river delta

1- 11 Rocks : 1 - Paleozoic ; 2 - Triassic ; 3 - Jurassic ;
4 - Jura - Cretaceous ; 5 - Mezo - Kainozoic granite ;
6 - Quaternary basalt ; 7 - Lower Pleistocene ;
8 - Middle Upper Pleistocene ; 9 - Upper Pleistocene
10 - Holocene ; 11 - Quaternary ;
12 - Hole position and the depth of bedrock (m)
13 - Isoline of thickness of Kainozoic deposits (m).

rface layer is of about 1-1.5 mm/year (vegetation vestiges with age C^{14} of 590 $\pm$ 150 years have been formed at the depth of 0.65m).

The climate characters of the peninsula are as follows (accor - ding to data acquired from meteorological station in Camau town) :

- Yearly total sunshine time : 2226 hours

- Yearly total radiation : 148.1 Kcalo/cm^2
- Yearly average temperature : 26.8 ^oC with Tmax = 28^oC and Tmin = 25.3 ^oC
- Yearly total rainfall : 2390 mm
- Yearly average relative humid : 85.9%
- East, north-east wind : from November to April of the next year
- South, south-west wind : from June to September

There is difference of tidal regimes between eastern and western seashores :

- At the eastern seashore : there is a semidayly inequality tidal regime with maximum amplitude of 3m

- At the western seashore : there is a diurnal inequality tidal regime with maximum amplitude of 1-1.5m

Because of favourable natural conditions such as low terrain , soil containing many nutritious substances, high dencity of chanal network and good climate as well as tidal conditions, the mangrove forest in this region is better growing and it occupies largest area in the country.

Methodology and Materials

The study is carried out based on the system and interdisciplinary approach combining with remote sensing. Some field trip profiles are carried out by visiting to affected areas.

There are essentially two main sources of information : those information from field investigation conducted as part of this study and those from previous related studies and past records. Aerial photographs acquired in 1968 (American AF-68), 1973 and 1991 for most important areas. The Landsat MSS acquired in 1973, 1982, 1986, 1989; the Thematic Maper data of Landsat 5 acquired in 1988, and 1990 were used to provide information of period from 1970 to present.

Some topographic and thematic maps at different scales are also used for this study.

Results and Discussion

The environmental changes

Intensive erosion process at the eastern seashore : During past 60 years, the length of eroded shoreline has been increased from 50 Km to 90 Km. This erosion process has moving tendency to the south - west direction with increasing eroded speed (Tab.1) /2/.

Accumudative process at the western seashore (in Camau bay):The assessment is carried out for 60 years period also. The obtained results show that the process is intensively going at this seashore and it can be showed at Tab.2 and Fig.2

Table 1 : Shoreline erosion at Camau peninsula's eastern seashore

Period	For the eastern seashore		For the Bode mouth
	Lossing area (Ha)	Eroded speed (m/year)	Eroded speed (m/year)
1930–1965	4,675	26	32
1965–1985	5,150	29	45.8

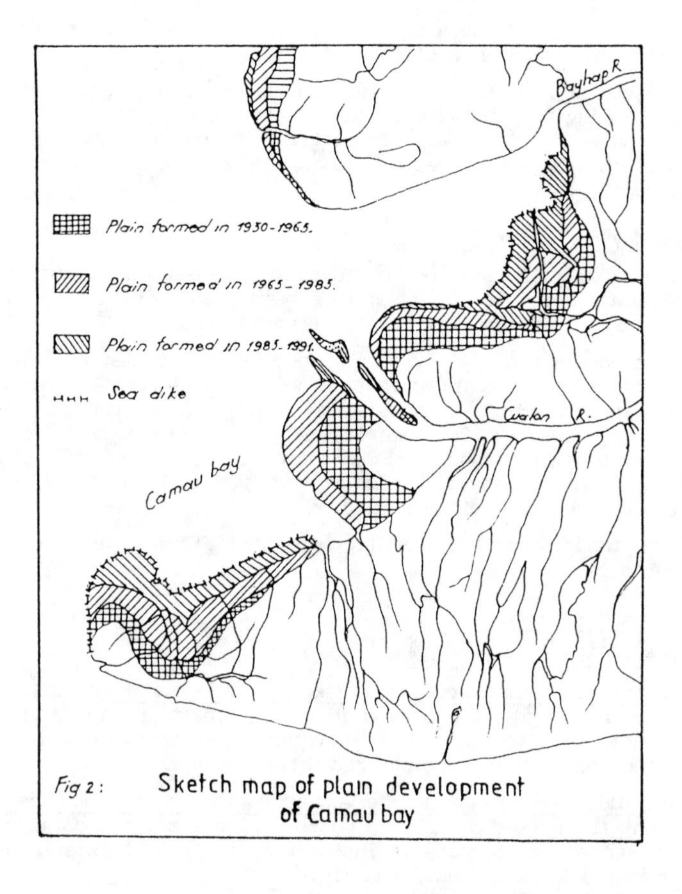

Fig 2: Sketch map of plain development of Camau bay

Table 2 : Accumudative process in the Camau bay /2/

Period	Alluvial area (Ha)	Alluvial speed (m/y)
1930-1965	3442	15.3
1965-1985	3410	26.6
1985-1991	1466	38.2

The mangrove forest area is fastly decreasing with its retrog - radation in quality (Tab.3)

Table 3 : Mangrove forest area in the Camau peninsula

Year	Mangrove forest area (Ha)
1943	200,000
1982	101,000
1988	75,000
1991	50,000

Land use change for the western area of the peninsula (Tab.4)

- Total deforested area by the war is 23.600 Ha and during following 13 years (1973-1986) is 38,900 Ha /3/

- The Tram (Melaleuca Leucadendron) forest area has also been fastly decreased with average speed of 5,852 Ha/year.

The main reasons for decreasion of forest area with high speed are followings :

- Population increases, especially mechanical increase.

- Deforestation for shirmp production without any managemental means.

Environmental impact assessment for dyke making for shirmp production in the Camau bay

The obtained research results /2/ show that the dyke making surrounding developing accumulative areas negatively influences on environment and natural resources in this area.

The change of geological-geomorphological environment

- The exchange process between sea and deposition areas is limited and due to mentioned the deposition speed is also reduced;those accumulative areas in back side of shirmp ponds receive inadequa-

Table 4 : Land use change in western area of Camau peninsula
(Ha)

Land use type	1969	1973	1979	1986
1. Water rice	232,800	252,500	253,600	243,600
2. Residential,upland crops and plantation	8,600	11,900	17,100	27,300
3. Commercial trees	3,300	-	-	800
4. Closed mangrove forest	74,900	51,300	39,000	12,400
5. Sparse and secondary mangrove forest	9,000	18,000	32,000	53,000
6. Closed Tram forest	130,000	48,300	31,400	13,500
7. Sparse and secondary Tram forest	1,800	61,600	71,200	59,800
8. Area for sea products production	-	200	2,200	9,200
9. Barren land with grass	43,600	65,000	69,300	79,600
10. Sediment beaches	1,100	9,400	2,400	600

te amount of sediments needed as usual.

- Those deposition areas in front of making dyke have not been developed due to lack of mangrove forest and surface erosion process

The change of hydrological and pedological environments

- The anuminous process of soil in ponds is intensively carried out.

- Water in ponds is degrading with tendency of anuminous increase and nourishing decrease.

- The soil of mangrove forest in back side of shirmp ponds is also degrading.

The change of biological society

- The development of vanguarded mangrove species (Avicennia) is hindered.

- The development of mangrove forest in back side of ponds is limited.

- The transportation of vegetation dregs from mangrove forest which are considered as foods for creatures living in shallow coastal areas is hindered.

The means for rational use of accumulative areas (tidal flood plain) in the CAmau bay

The principles must be to consider when use accumulative areas

- To ensure the closed relationship between sea and mangrove fo rest. A regular amount of sediment from the sea is needed for the forest from which vegetation dregs could be transported to the sea.

- To ensure encroached process of mangrove forest on the sea

- To ensure relationship between eastern and western seashores, so eroded products of eastern seashore could be transported to the western one through chanal network.

The line of uses

- To protect those mangrove forest that are encroaching on the sea.

- To combine forest growing with sea product making on new formed accumulative area as well as on forming one.

- To use limited area behind protected forest for improved shirmp ponds with high productivity.

- To carry out as soos as possible the concrate agriculture - forestry-fishing planning development project for the hole mangrove forest area in the Camau peninsula.

References

1. Le Duc An, 1982. Morphostructural features of the lower Mekong plain. Journal " Geomorphology " N4, Moscow. (p.87-93).

2. Le Duc An, Pham Trung Luong, 1992. The geoenvironment of the playas in Camau gulf and problems of changes due to human impact. Proceeding of Regional Seminar on Environmental Geology,Hanoi.(p,198 -203)

3. Pham Trung Luong, 1992. Application of remote sensing to land use studies in Vietnam. Doctorate thesis of Geography-Geological sciences, Hanoi, (123p.) (in Vietnamese).

Coastal processes and management of the Mahanadi river deltaic complex, East coast of India

Manmohan Mohanti[1]

Abstract

The Mahanadi river deltaic complex in the tropical climatic setting on the east coast of India has been studied with respect to geomorphology, land use/cover, coastal processes and management. The coast is prone to natural hazards like river floods and cyclones. Littoral drift of sands in order of 1.5 million tons move annually from southwest to northeast. Dynamic processes are monsoon-dominated as well as episodic event-dominated. Shore erosion, wetland loss as well as other anthropogenic activities cause concern. Management controls are sought for proper land and water use, checking flood havoc, conservation of mangrove resources and protection from shore erosion.

Introduction

River deltas resulting from interacting fluvial and marine forces exist in a wide spectrum of settings and are of immense benefit to the people inhabiting its fertile, low-lying terrain. Deltas of different geologic, geographic and climatic settings have their own characteristics and vary in terms of morphologic suites, overall geometry, sediment properties and dynamic environment (Wright, 1985). Patterns of land use/cover vary in deltaic complexes depending upon terrain setting and human settlements flourishing on the deltaic plains. Many present-day deltas are experiencing large coastal landloss due to geological, catastrophic, biological and man-induced factors (Coleman and Roberts, 1989). For sustainable development of the deltaic resources and restoration of environments and ecosystem, increasing attention is now paid to conservation and management programmes.

---

[1]Senior Reader, Department of Geology, Utkal University, Vani Vihar, Bhubaneswar-751004, India.

In India modern deltas are particularly conspicuous on the east coast on the Bay of Bengal. Studies on diverse aspects of the Mahanadi river deltaic complex, one of such deltas on the east coast add further informations to the deltaic studies on the whole. Informations are presently focussed on geomorphology, coastal processes and management programme of the Mahanadi River deltaic complex, in the Cuttack District north of the Devi river distributary.

Location and climate

The Mahanadi river following ca. 851 km long course drains a catchment area of 1.42×10^5 km^2 and forms an arcuate delta before debouching into the Bay of Bengal. The delta plain of 0.9×10^4 km^2 lies between 85° 40' : 86°45'E and 19° 40' : 20° 35'N and covers parts of Cuttack and Puri Districts, Orissa State. The river after entering the deltaic plains at Naraj branches off and develops its distributary system with Birupa river lying on the northern side and Daya river on the southern side (Fig.1).

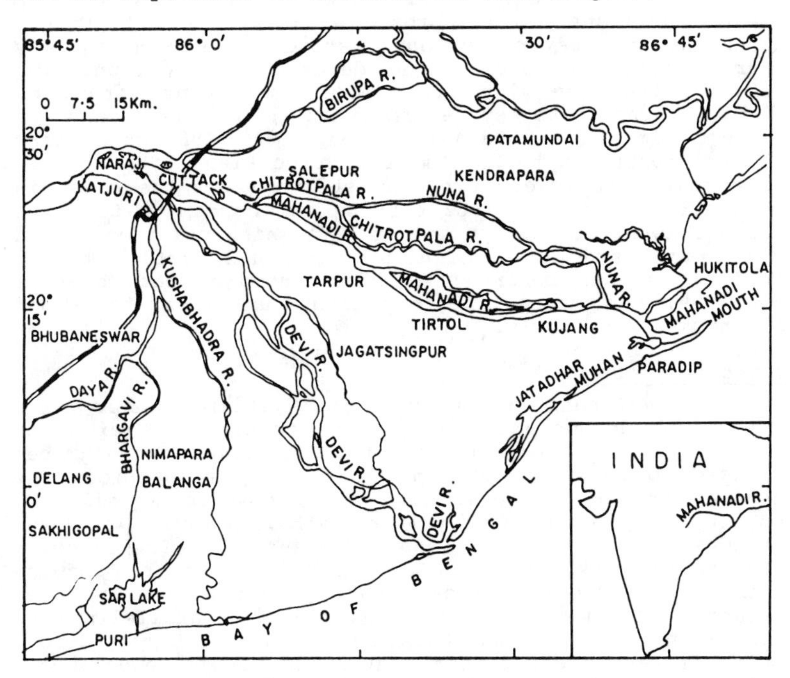

Fig.1.Mahanadi river deltaic distributary system.

The climate of the area is tropical monsoon climate, hot and humid. The average annual rainfall amounts to ca.1572 mm out of which ca. 73% is precipitated during the vigorous southwest monsoon between mid. June to mid. October and ca. 17 o/o is dropped during the milder northeast monsoon period between December to early January. Remaining 10 o/o is scattered during rest of the year. Temperature may rise upto 44°C or even more during summer months of May and June. Temperature varies between 12°C-35°C during winter months between November to February. Two dominant wind patterns prevail; a southwesterly wind during April - October and a northeasterly wind during November-January.

The coastal tract is prone to tropical depressions/cyclones during the pre-monsoon months of May-June and post-monsoon months of October-November when depressions may arise in the Bay of Bengal leading to cyclonic storms of severity. Severe tropical storms are marked by wind speed of 89-116 km/hr and severe storm with core of hurricane winds approach wind speed of 118 km/hr. Consequent upon severe cyclones, storm surge 6-7m high may enter into the coast and cause devastations by inundating parts of the low-lying areas with subsequent spread of salinity. The severest storm that hit this coast in recorded time was at Paradip in September 1885 when storm velocity mounted upto 250 km/hr (Action Plan, 1988-91). In the recent past severe cyclonic storms swept the coast in October 1971 and June 1982 causing severe damage to parts of the coastal tract. Cyclone of 1971 which was accompanied by a storm surge of 6-7m high over considerable stretch north of Paradip reported to have caused a death toll of about 10,000 persons besides enormous loss to other livestock and property.

Geological setting

The Mahanadi Basin with its deltaic setting was formed in a downwarp on the passive continental margin of the east coast of India forming part of the Gondwana graben believed to be a failed arm of the triple junction. The onshore part of the basin is mostly covered by alluvial deposits of the Mahanadi river delta. Towards west and northwest of Cuttack and Bhubaneswar Pre-Cambrian metamorphic rocks of the Eastern Ghats, Upper Gondwana sedimentaries and laterites occur. Mahanadi offshore area appears to be a pull-apart type of basin paralleling the coast with rapid seaward thickening of sedimentary section. Faults present are basement controlled penecontemporaneous faults and also some growth faults (Jagannathan et al., 1983). Deep Seismic Profiling onshore revealed the basement configuration to be uneven with ridges and depressions affected by faulting and probably movements are now also controlled along the faults. The depressions have sediment thickness over 2000m implying continued subsidence (Baishya and Singh, 1985).

Various landforms in the deltaic complex have been interpreted from LANDSAT-5 (TM) data and depicted in the geomorphological maps (Figs.2a, b; Fig.3). The landforms have been grouped under Alluvial Flood Plan, Coastal Plain and Tidally Influenced categories. Offshore data suggest major delta building started during Middle Miocene (Bherali et al.,1992). In course of the Holocene evolution of the Mahanadi delta a complex of barrier islands - spits have developed and there has been progradation of the coast in a northerly direction.

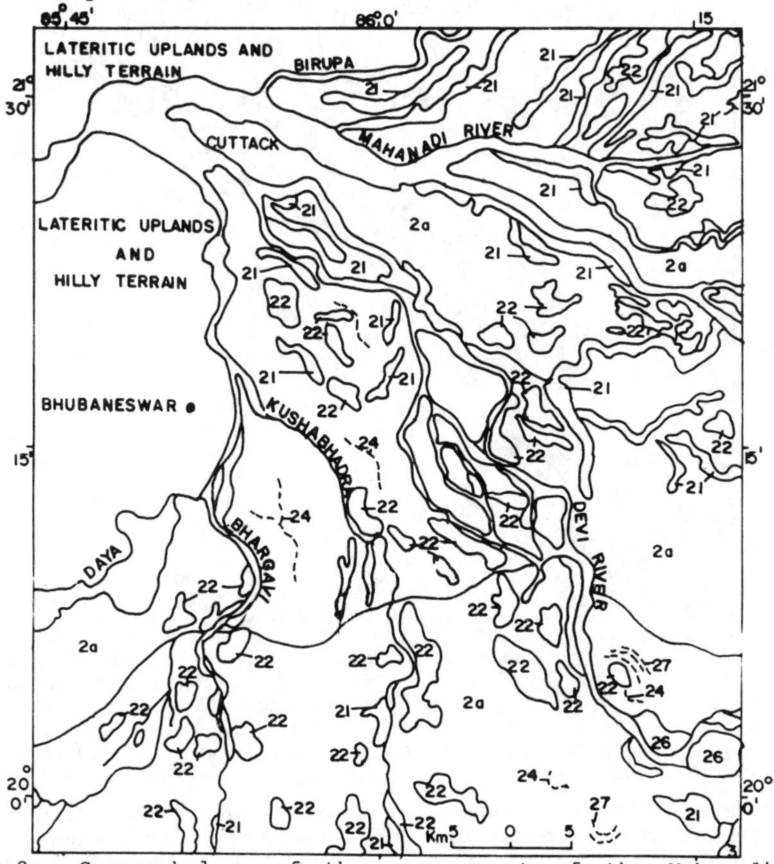

Fig.2a. Geomorphology of the upper part of the Mahanadi deltaic complex interpreted from LANDSAT-5 (TM) data, 9th Dec.1985 (Courtesy : ORSAC). 1.Lateritic uplands and hilly terrain, 2.Alluvial flood plain : 2a - Upper flood plain, 21-Natural levee, 22-Backswamp, 24-Abandoned channel, 26-Channel bar, 27-Meander scar.

Fig.2b.(in continuation of Fig.2a). Geomorphology of the lower part of the Mahanadi deltaic complex. 2.Alluvial flood plain : 2b-Lower flood plain, 23-Palaeo channel,25-Point bar, 3.Coastal plain : 31-Beach, 32-Sand dune, 33-Spit, 34-Beach ridge, 35-Swale, 36-Barrier island, 37-Shoal, 4.Tidally influenced group : 41-Mangrove, 42-Tidal creek, 43-Intertidal flat, 44-Tidal swamp.

Meijerink (1982-83) traced shift of Mahanadi mouth and had drawn attention to position of shorelines close to Hukitola barrier island over a period of 150 years. Published informations (Sambasiva Rao et al.,1978; Chakraborty and Chattopadhyay, 1989) on beach ridges marking former strandlines and discussed in their geomorphological context and C^{14} dating data suggest major force of delta building spanned during the interval 6000 to 800 Yr.B.P. Further seaward shifting of shoreline has also occurred during last 800 years.

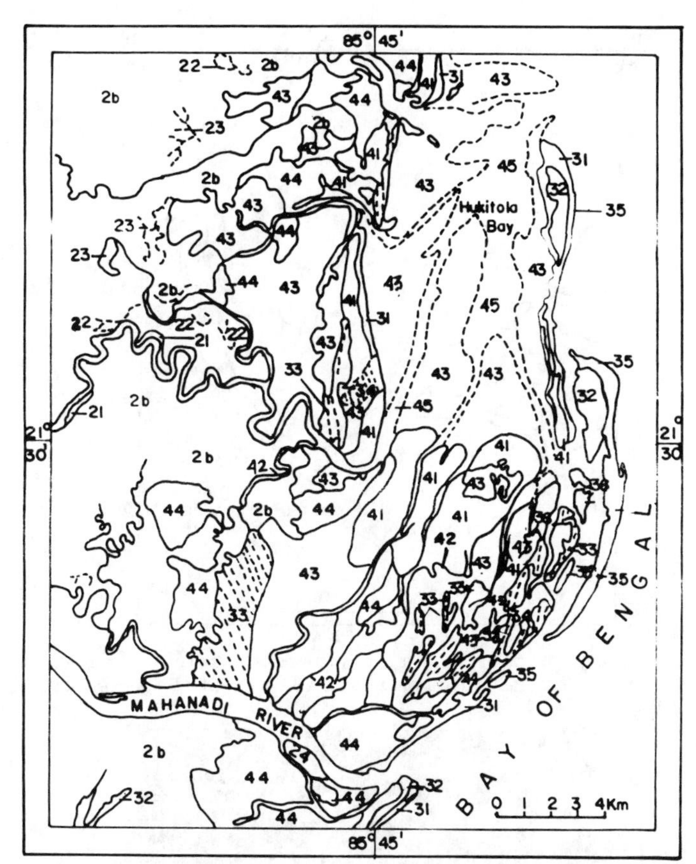

Fig.3. Details of geomorphology north of the Mahanadi river mouth in the Hukitola Bay region interpreted from LANDSAT-5 (TM) data, 9th Dec.1985 (Courtesy ORSAC). 2.Alluvial flood plain : 2b-Lower flood plain, 21-Natural levee, 22-Palaeochannel, 23-Abandoned channel, 24-Channel bar, 3.Coastal plain : 31-Beach, 32-Sand dune, 33-Beach ridge, 34-Swale, 35-Barrier island, 36-Shoal, 4.Tidally influenced group : 41-Mangrove, 42-Tidal creek, 43-Intertidal flat, 44-Tidal swamp, 45-Subtidal flat.

A barrier spit which was present on the southwest side of Mahanadi mouth (Fig.4) and had grown in the northeasterly direction at approx. rate 500m per year between 1962-1979 (remarkably faster after the construction of breakwaters for Paradip Port in the later part of sixties) was finally destroyed after the heavy floods of August 1982, thus shifting the position of the Mahanadi mouth (Fig.5).

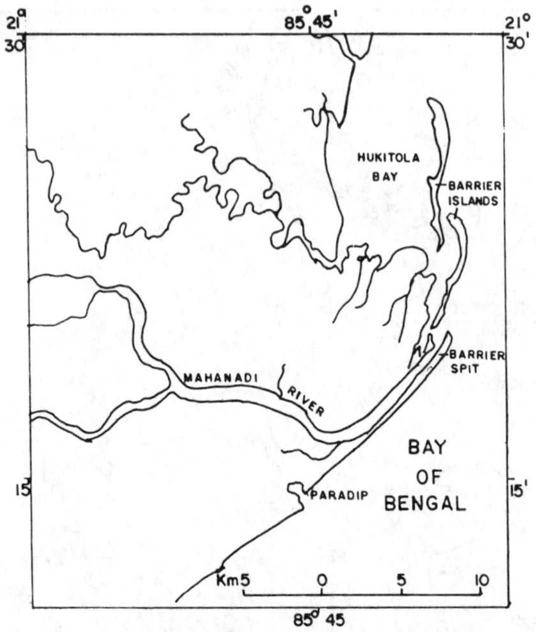

Fig.4. Barrier-spit on the southwestern side of the Mahanadi river mouth interpreted from LANDSAT-1, MSS Band-7, 1975, (Courtesy : ORSAC).

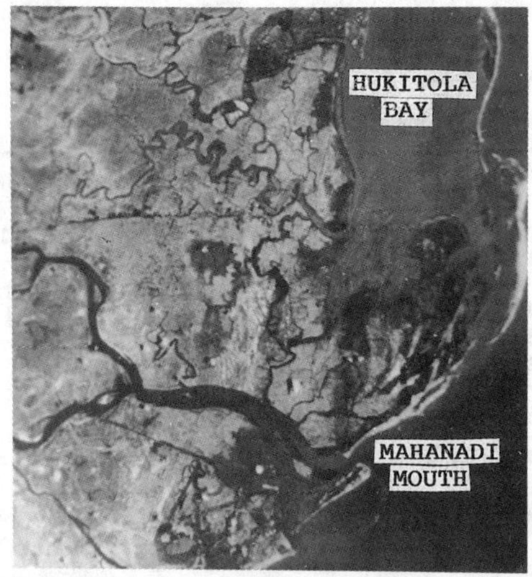

Fig.5. Hukitola Bay region. Barrier-spit (Fig.4) is destroyed. IRS-1A, Liss-2 imagery, 19th April, 1988, (Courtesy : ORSAC).

Coastal processes

The deltaic coast is microtidal, monsoonal (southwest monsoon) wave-dominated to mixed wave-tide-dominated. In the Mahanadi river deltaic distributary system, the Mahanadi river channel proper and the Devi river are the two prominent estuaries. The estuaries are microtidal, mean tidal range near Mahanadi mouth being 1.29m with semi-diurnal tidal cycles. The width of the shelf at Paradip is ca. 31.5 km and shelf break occurs in a water depth of ca. 47m. The siliciclastic shore faces dominant wave action during southwest monsoon when swell waves due to southwesterly monsoonal winds obliquely strike the shore from the southwest with wave heights reaching 3m or even more. Wave direction changes from southwest to northeast during relatively quieter post-monsoonal month of late October-November and milder northeast monsoon period of December to early January when wave height is reduced to about 1m or less. Surface current pattern on the shelf is more complex, however, the current strength may be generally reaching upto 0.53m/sec. The monsoonal fluvial discharge into the Bay of Bengal is remarkable. The Mahanadi river at its delta head at Naraj carries an annual average discharge (1975-1984) of water estimated to be 48,691 million cubic metres out of which the average monsoonal component during July-September (1980-1984) was estimated to be 41,000 million cubic metres by the Irrigation Department, Govt. of Orissa. During monsoonal floods fluvial discharge is enhanced. A water discharge of 17,150 cumsecs initiates flood in the Mahanadi river. Damaging floods may occur in an interval of 3-5 years when water discharge rises to 28,580 cumsecs. It is worth mentioning that the severest flood of the century passed in the Mahanadi river within two consecutive days in August-September 1982 when water discharge mounted upto 44,749 cumsecs causing unprecedented devastation in the deltaic plains.(Fig.6).

Monsoonal processes dominate and involve highest fluvial water discharge and suspended sediment input (99.96%) (Ray, 1988) into the Bay of Bengal with a high flushing velocity. With low volumetric capacity of the river and high flushing velocity, the suspended load enters into the Bay of Bengal probably as a hypopycnal buoyant plume and/or friction-dominated plane jet depending upon fluvial discharge and nearshore hydrographic regime. Fine particles may be dispersed as turbid plumes seen at the river mouth (Fig.7). The delta head receives an average annual suspended sediment load of 27.07 million tons and bed load of 2.70 million tons during monsoonal months (July-September). Sands in order of 75,000 to 150,000 tons/yr. may be reaching the shore only through the Mahanadi river mouth after deposition in parts of flood

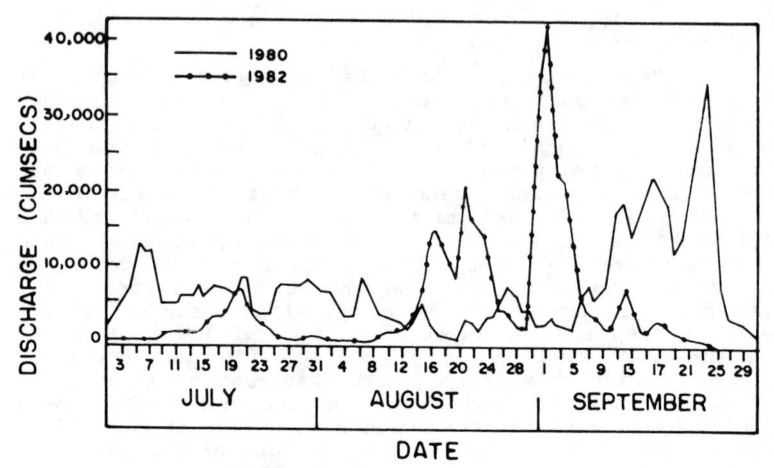

Fig.6. Daily water discharge (southwest monsoonal) through the Mahanadi river at its delta head at Naraj. Highest flood water discharge occurred during Aug.31-Sept.1, 1982 (Data source : Irrigation Department, Govt.of Orissa Courtesy).

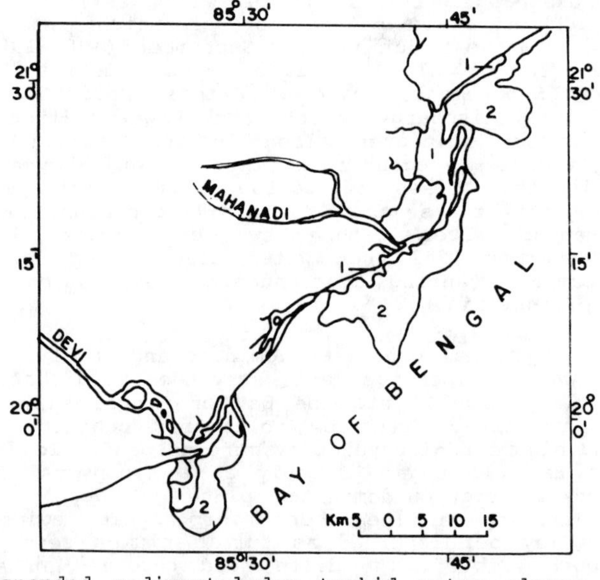

Fig.7.Suspended sediment-laden turbid water plumes at the river mouths interpreted from Indian Remote Sensing Satellite (IRS-1A,Liss-1) data, 12th Oct.,1988. High turbid water-1, Relatively low turbid water-2(Courtesy : ORSAC).

plain, channel and tidal areas (Meijerink, 1982-83). During southwest monsoon waves breaking obliquely to the shore generate littoral drift of sands moving parallel to the shore from southwest to northeast. A littoral drift in order of 1.5 million tons move annually along Paradip coast. The monsoonal fluvial transport of sands and wave induced northerly littoral drift build up a depocentre for sands in the nearshore region shaping and supporting a complex of barrier island-bar-spit system in the vicinity of the Mahanadi river mouth. Offshore region, mangrove swamps and tidal creek act as sinks for fine sediments (Ray and Mohanti, 1989). In the relatively quieter non-monsoonal period fine sediments may be distributed by tidal action and residual estuarine circulation with some possible supply from offshore. Landward transport of bottom sediments may be taking place by relatively stronger flood tidal currents during pre-monsoon months. The sediment depositional pattern on the deltaic coast may be altered considerably by episodic events like damaging floods and severe storms/cyclones occurring in the coastal area. Severe floods could cause destruction of varying magnitude in the barrier-spit system near the river mouth. Formation and destruction of sand-spit is a recurring phenomenon. A breach may also occur during an exceptionally violent storm. Storm erosion may add material to the littoral drift.

Management of the deltaic complex

The delta is thickly populated (approx. 3×10^6 people), the chief occupations of the people being agriculture and fishing. However, with growing population pressure and uncontrolled anthropogenic activities sustainable development and management programmes are warranted for arresting environmental degradation and raising resource potential in the delta area. The land use/cover map is given in Fig.8a and b. Management of the delta complex urgently should take into consideration the land and water use, mangrove conservation and shore erosion controls.

Major problems with regard to land and water management are : (a) drainage congestion and waterlogging, (b) detrimental tidal ingress and inundation particularly during storms and subsequent spread of salinity, (c) flood havoc and (d) uncontrolled anthropogenic activities. Drainage congestion and waterlogging are conspicuous in interfluve areas of middle and lower parts of the deltaic plains. There is heavy growth of aquatic plants like water hyacinths in stagnant waterlogged bodies due to lack of adequate drainage and rise of water table. Modern irrigation programmes will improve this situation. The principal paddy cultivation being monsoon dependant, the ill effects of the vagaries of monsoon can be mitigated to a large extent by modernising irrigation system taking into consideration of alternate strategies of crop production,canal regulation and conjunctive use of

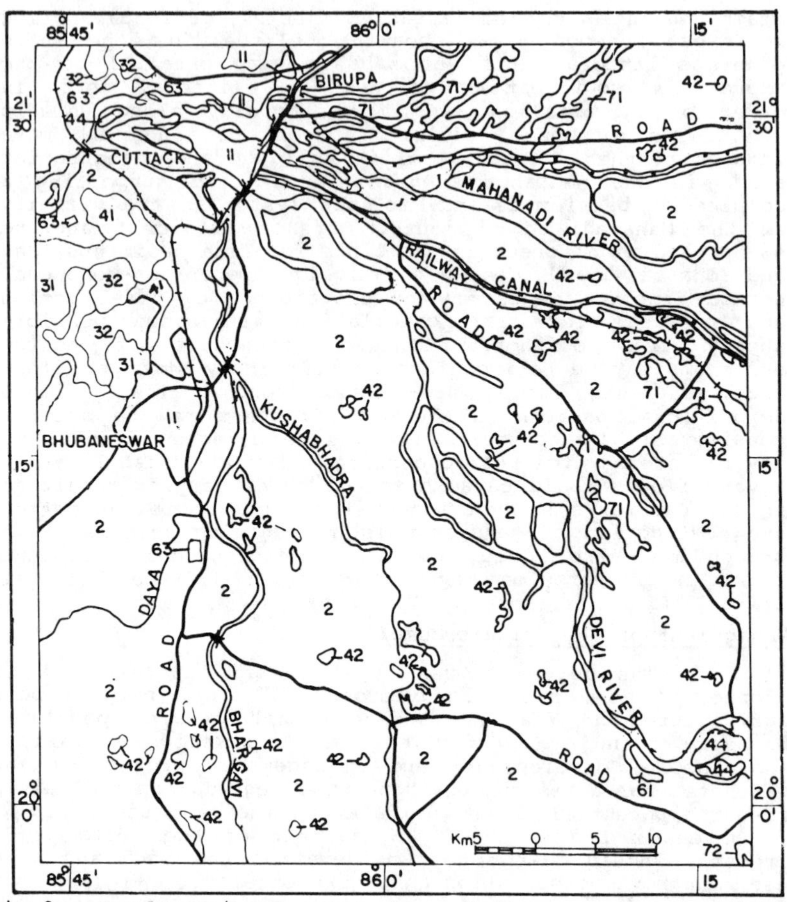

Fig.8a. Land use/Land cover map of the upper part of the Mahanadi deltaic complex interpreted from Indian Remote Sensing Satellite (IRS-1A, Liss-1) data, 21st Feb. 1989 (Courtesy : ORSAC). (Legend of Fig.8a and 8b combined). 1.Built-up-land : 11-Settlements (include partly industrial areas, only city settlements are shown), 12-Industrial area, 2-Agricultural land, 3.Forest : 31-Dense/Sparse, 32-Degraded, 33-Mangrove, 34-Forest plantation, 4. Wasteland : 41-Undulating upland with or without scrubs, 42-Waterlogged land, 43-Sandy area with vegetation, 44-Sandy area without vegetation, 5.Coastal Wetland : 51-Mudflat, 52-Tidal Swamp, 6. Waterbodies : 61-River/Stream/Creek, 62-Canal(numbers are not shown on the map as they are self explanatory), 63-Reservoir/Ponds/Tanks,7.Others : 71-Settlements with mixed vegetation, 72-Salt pan.

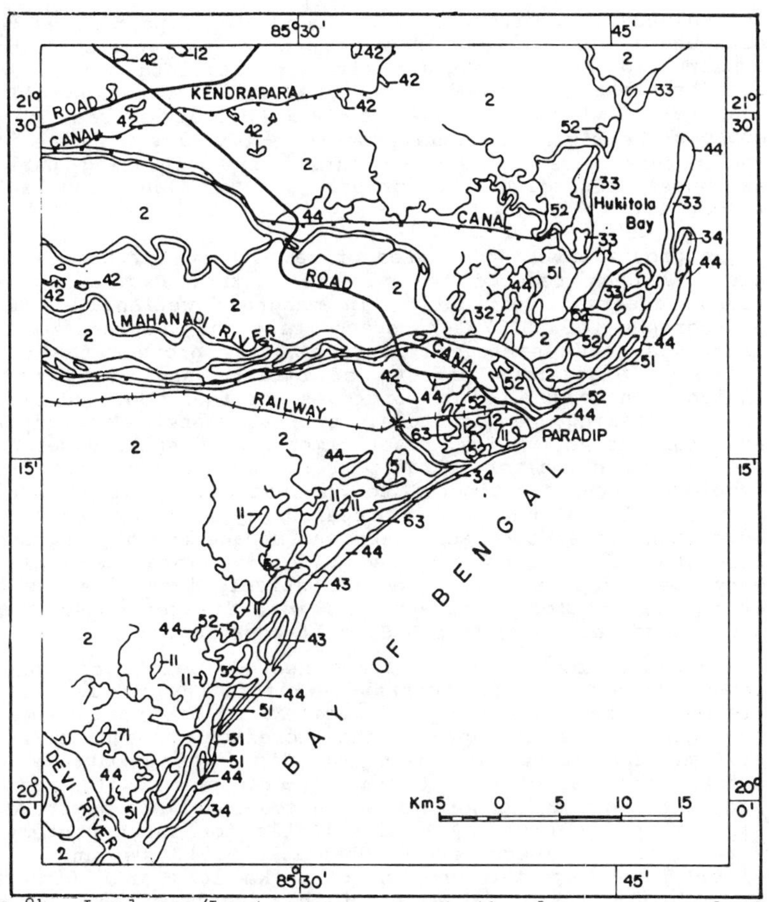

Fig.8b. Land use/Land cover map of the lower part of the Mahanadi deltaic complex.

Groundwater with surface canal water flow from the river system. Salinity problems in coastal aquifers vary with depth as tidal effects and salt water intrusion are observed 20-60 km upstream from the shoreline. In the management of coastal groundwater for maintenance of freshwater aquifers for usable purposes, the natural extent of saltwater wedge in the aquifer or the upconing of salt water due to excess ground water development should be studied in depth with periodic monitoring (Das, 1992). The Mahanadi Delta Command Area master plan (Irrigation Dept., Govt. of Orissa) is going to be implemented for the optimal use of water and land. Tidal ingress in estuarine tracts ca. 35 km, inland, gets further accentuated under severe storms. Saline embankments and other suitable engineering constructions are deemed necessary with proper environmental impact assessment. Although a major dam, a weir and a barrage have been constructed on the Mahanadi river to regulate water flow, apprehensions of flood disasters exist and controlling measures need to be taken in the deltaic areas during monsoon. Detrimental anthropogenic activities interfering the natural system must be curbed by generating public awareness regarding the proper use of land and water resources.

Mangroves covering an area of ca. 200 sq.kms.before 1940 has now been reduced to only 50 sq.kms. Parihar et al. (1986) observed degradation in mangrove wetland in TERRA photographs of SALYUT-7 space mission. Degradation of mangroves cause serious concern as they protect the coast from the disastrous effects of cyclone and storm surge. Indiscriminate cutting of mangroves for fuel wood and timber cattle grazing, agricultural encroachments, unscientific acquaculture activities, resettlements of displaced persons, and construction of Paradip Port and other industrial complexes have contributed to the wetland loss. An Action Plan (1988-91) has been prepared by the Orissa State Government towards conservation and management of mangroves. Banerjee (1992) also stressed effective management plans. Mangroves in Orissa has been tentatively identified as one of the five selected eco-systems in India for conservation and management under Indo-U.S. collaboration.

Shore erosion has been marked near Paradip Port (due to construction of breakwaters) and Hukitola Bay barrier island region. Shore erosion may be accelerated as there seems to be reduced sediment input to the shoreface through the river systems due to construction of dam, weir, barrage and embankments arresting sediment transport. If reduced sediment input and general rise in sea-level are admitted, a staggering landloss may be inevitable for which protection strategies are necessary (Mohanti, 1990). De and Ghosh (1981) have reported erosion of 5 km long and 300m wide stretch of deltaic coastal tract in a period of 50 years. In addition to existing shelter belt plantations of casurina and cashew nuts in parts of sandy coastal stretches, construction of sand fences and plantation of dune grasses and marsh vegetation may be implemented (Savage and Woodhouse, 1968; Penland and Suter, 1988) to arrest erosive processes apart from shoreface/beach nourishment programmes.

Concluding remarks

The fertile deltaic complex of the Mahanadi river with bountiful agricultural, fishery and mangrove resource potential and having touristic, aesthetic, ecological and economic importance need to be saved from environmental degradations through appropriate conservation and management programmes. A Coastal Zone Management Plan is being developed at the Orissa Remote Sensing Application Centre (ORSAC).

Acknowledgements :- I am thankful to P.R.Mohanty,IFS,Chief Executive, ORSAC for the facility of using the imageries & P.Kumar,Jr.Scientist for discussion.

Appendix

References

Action Plan for protection and propagation of mangroves in Mahanadi Delta of Kujang Range (From 1988-89 to 1990-91). Divisional Forest Officer, Wildlife Conservation Division, Chandbali, Balasore (Orissa), India, 51 p.

Baishya, N.C. and Singh, S.N. (1985) DSS profiling in Mahanadi onshore areas : A case study, in Proc. Int. Symp. on Deep Seismic Sounding Traverses. Ed. by K.L.Kaila and H.C. Tewari, Assoc. Explor. Geophys., Hyderabad, India, pp. 1-9.

Banerjee, L.K. (1992) Mangal Formations of the Mahanadi Delta : Exploitation and Management, in Tropical Ecosystems : Ecology and Management. Ed. by K.P.Singh and J.S.Singh., Wiley Eastern Ltd., New Delhi, pp. 289-294.

Bharali, B., Rath, S. and Sarma, R. (1992) A brief review of Mahanadi Delta and the deltaic sediments in Mahanadi Basin, in Quaternary Deltas of India. Ed. by R.Vaidyanadhan, Geol. Soc. India, Memoir No.22, pp. 31-49.

Chakraborty, S.C. and Chattopadhyay, G.S. (1989) Quaternary Geology and Geomorphology of the Mahanadi wave dominated delta in Orissa. Abstract Volume, Workshop on Coastal Processes and Coastal Quaternaries of Eastern India, Geol. Surv. India (Eastern Region, Calcutta), pp. 9-11.

Coleman, J.M. and Roberts, H.H. (1989) Deltaic coastal wetlands, in Coastal Lowlands : Geology and Geotechnology. Ed. by W.J.M. Van der Linden, S.A.P.L. Cloetingh, J.P.K. Kaasschieter, W.J.E. van de Graaff, J. Vandenberghe and J.A.M. van der Gun. Proc. Symp. Royal Geol. and Mining Soc., Netherlands, Kluwer Acad. Publ., Dordrecht, pp. 1-24.

Das, S. (1992) Hydrogeological features of Deltas and Estuarine tracts of India, in Quaternary Deltas of India. Ed. by R.Vaidyanadhan, Geol. Soc. India, Memoir No.22, pp. 183-225.

De, S.K. and Ghosh, R.N. (1981) Report on the geomorphological and Quaternary geological investigations around Mahanadi Delta in parts of Puri and Cuttack Districts, Orissa. Geol. Surv. India (Eastern Region, Calcutta) unpublished Prog. Rep. for field season 1978-79.

Jagannathan, C.R., Ratnam, C., Baishya, N.C. and Das Gupta, U. (1983) Geology of the offshore Mahanadi Basin. Petroleum Asia Jour., Vol.4 (4), pp.101-104.

Meijerink, A.M.J. (1982-83) Dynamic geomorphology of the Mahanadi Delta. ITC Jour., Verstappen Volume, Enschde, Netherlands, pp.243-250.

Mohanti, M. (1990) Sea Level Rise : Background, Global Concern and Implications for Orissa Coast, India, in Sea Level Variation and its Impact on Coastal Environment. Ed. by G. Victor Rajamanickam, Tamil University, Publ. No.131, Thanjavur, India, pp. 197-232.

Parihar, J.S., Panigrahy, S. and Das, K.C. (1986) Environmental assessment of parts of Mahanadi and Brahmani River Deltas of Orissa using Salyut-7 Space photographs. Results from the Joint Indo-Soviet Remote Sensing Experiment Terra on board Salyut-7, Spec. Publ., ISRO-SP-17-86, Bangalore, India, pp. 21-31.

Penland, S. and Suter, J.R. (1988) Barrier Island erosion and protection in Louisiana : A coastal geomorphological perspective. Trans. Gulf Coast Assoc. Geol. Soc., Vol. 38, pp.331-342.

Ray, S.B. (1988) Sedimentological and Geochemical studies on the Mahanadi River estuary, East Coast of India. Unpublished Ph.D. thesis, Utkal University, Bhubaneswar, India, 204 p.

Ray, S.B. and Mohanti, M. (1989) Sedimentary processes in the Mahanadi River estuary, East Coast of India. Abstract Volume, Workshop on Coastal Processes and Coastal Quaternaries of Eastern India, Geol. Surv. India (Eastern Region, Calcutta) pp.9-11.

Sambasiva Rao, M., Nageswara Rao, K. and Vaidyanadhan, R. (1978) Morphology and evolution of Mahanadi and Brahmani-Baitarani Deltas. Proc. Symp. Morphology and evolution of Landforms. Dept. of Geology, University of Delhi, pp.241-249.

Savage, R.P. and Woodhouse, W.W.Jr. (1968) Creation and Stabilization of coastal barrier dunes. Proc. 11th Conf. Coastal Engineering, London, England [Reprint 3-69, September 1968, U.S.Army Coastal Engineering Research Center], pp.671-700.

Wright, L.D. (1985) River Deltas, in Coastal Sedimentary Environments. Ed.by Richard, A. Davis, Jr., Springer-Verlag, New York Inc., pp.1-76.

SUBJECT INDEX
Page number refers to first page of paper

AUTHOR INDEX
Page number refers to first page of paper